INVESTIGATIONS IN NUMBER, DATA

Developing Number Sense

Collecting, Counting, and Measuring

Kindergarten

Megan Murray
Karen Economopoulos
Marlene Kliman

Developed at TERC, Cambridge, Massachusetts

Dale Seymour Publications®
White Plains, New York

The *Investigations* curriculum was developed at TERC (formerly Technical Education Research Centers) in collaboration with Kent State University and the State University of New York at Buffalo. The work was supported in part by National Science Foundation Grant No. ESI-9050210. TERC is a nonprofit company working to improve mathematics and science education. TERC is located at 2067 Massachusetts Avenue, Cambridge, MA 02140.

This project was supported, in part, by the
National Science Foundation
Opinions expressed are those of the authors and not necessarily those of the Foundation

Managing Editor: Catherine Anderson
Series Editor: Beverly Cory
Manuscript Editor: Nancy Tune
ESL Consultant: Nancy Sokol Green
Production/Manufacturing Director: Janet Yearian
Production/Manufacturing Manager: Karen Edmonds
Production/Manufacturing Coordinators: Joe Conte, Roxanne Knoll
Design Manager: Jeff Kelly
Design: Don Taka
Composition: Archetype Book Composition
Illustrations: Rachel Gage, DJ Simison, Susan Jaekel
Cover: Bay Graphics

This book is published by Dale Seymour Publications®, an imprint of Addison Wesley Longman, Inc.

    Dale Seymour Publications
    10 Bank Street
    White Plains, NY 10602
    Customer Service: 1-800-872-1100

Copyright © 1998 by Dale Seymour Publications®. All rights reserved. Printed in the United States of America.

Limited reproduction permission: The publisher grants permission to individual teachers who have purchased this book to reproduce the blackline masters as needed for use with their own students. Reproduction for an entire school or school district or for commercial use is prohibited.

Order number DS47105
ISBN 1-57232-928-9
8 9 10-ML-02 01

Printed on Recycled Paper

TERC

INVESTIGATIONS IN NUMBER, DATA, AND SPACE®

Principal Investigator Susan Jo Russell
Co-Principal Investigator Cornelia Tierney
Director of Research and Evaluation Jan Mokros
Director of K–2 Curriculum Karen Economopoulos

Curriculum Development
Karen Economopoulos
Rebeka Eston
Marlene Kliman
Christopher Mainhart
Jan Mokros
Megan Murray
Kim O'Neil
Susan Jo Russell
Tracey Wright

Evaluation and Assessment
Mary Berle-Carman
Jan Mokros
Andee Rubin

Teacher Support
Irene Baker
Megan Murray
Kim O'Neil
Judy Storeygard
Tracey Wright

Technology Development
Michael T. Battista
Douglas H. Clements
Julie Sarama

Video Production
David A. Smith
Judy Storeygard

Administration and Production
Irene Baker
Amy Catlin

Cooperating Classrooms for This Unit
Jeanne Wall
Arlington Public Schools
Arlington, MA

Audrey Barzey
Patricia Kelliher
Ellen Tait
Boston Public Schools
Boston, MA

Meg Bruton
Fayerweather Street School
Cambridge, MA

Rebeka Eston
Lincoln Public Schools
Lincoln, MA

Lila Austin
The Atrium School
Watertown, MA

Christopher Mainhart
Westwood Public Schools
Westwood, MA

Consultants and Advisors
Deborah Lowenberg Ball
Michael T. Battista
Marilyn Burns
Douglas H. Clements
Ann Grady

CONTENTS

About the *Investigations* Curriculum	I-1
How to Use This Book	I-2
Technology in the Curriculum	I-6
About Assessment	I-8

Collecting, Counting, and Measuring

Unit Overview	I-10
Materials List	I-17
About the Mathematics in This Unit	I-18
About the Assessment in This Unit	I-20
Preview for the Linguistically Diverse Classroom	I-22

Investigation 1: Counting Books	2
Investigation 2: Taking Inventory	22
Investigation 3: Comparing Towers	36
Investigation 4: Counting and Comparing	52
Investigation 5: Least to Most	66
Investigation 6: Arrangements of Six	78

Choice Time Activities

My Counting Book	10
Grab and Count	12
Counting Jar	14
Inventory Bags	30
Measuring Table	42
Grab and Count: Which Has More?	44
Compare	46
Comparing Names	60
Grab and Count: Compare	62
Collect 10 Together	64
Grab and Count: Least to Most	72
Racing Bears	74
Books of Six	86

General Teacher Notes	92
About Classroom Routines	99
Tips for the Linguistically Diverse Classroom	110
Vocabulary Support for Second-Language Learners	112
Blackline Masters: Family Letter, Student Sheets, Teaching Resources	113

TEACHER NOTES

Collaborating with the Authors	I-20
Counting Is More Than 1, 2, 3	16
Students' Counting Books	19
Observing Students As They Count	32
From the Classroom: Students' Difficulties with Counting	34
Learning About Length	50
Why Six?	88
From the Classroom: "I Had One Too Much"	89
Grab and Count and Its Variations	91
About Choice Time	92
Materials as Tools for Learning	95
Encouraging Students to Think, Reason, and Share Ideas	96
Games: The Importance of Playing More Than Once	97

WHERE TO START

The first-time user of *Collecting, Counting, and Measuring* should read the following:

- About the Mathematics in This Unit — I-18
- About the Assessment in This Unit — I-20
- Teacher Note: Collaborating with the Authors — I-20
- Teacher Note: Counting Is More Than 1, 2, 3 — 16
- Teacher Note: Observing Students As They Count — 32
- Teacher Note: Learning About Length — 50
- Teacher Note: About Choice Time — 92

When you next teach this same unit, you can begin to read more of the background. Each time you present the unit, you will learn more about how your students understand the mathematical ideas.

ABOUT THE *INVESTIGATIONS* CURRICULUM

Investigations in Number, Data, and Space® is a K–5 mathematics curriculum with four major goals:

- to offer students meaningful mathematical problems
- to emphasize depth in mathematical thinking rather than superficial exposure to a series of fragmented topics
- to communicate mathematics content and pedagogy to teachers
- to substantially expand the pool of mathematically literate students

The *Investigations* curriculum embodies a new approach based on years of research about how children learn mathematics. Each grade level consists of a set of separate units, each offering 2–8 weeks of work. These units of study are presented through investigations that involve students in the exploration of major mathematical ideas.

Approaching the mathematics content through investigations helps students develop flexibility and confidence in approaching problems, fluency in using mathematical skills and tools to solve problems, and proficiency in evaluating their solutions. Students also build a repertoire of ways to communicate about their mathematical thinking, while their enjoyment and appreciation of mathematics grows.

The investigations are carefully designed to invite all students into mathematics—girls and boys, members of diverse cultural, ethnic, and language groups, and students with different strengths and interests. Problem contexts often call on students to share experiences from their family, culture, or community. The curriculum eliminates barriers—such as work in isolation from peers, or emphasis on speed and memorization—that exclude some students from participating successfully in mathematics. The following aspects of the curriculum ensure that all students are included in significant mathematics learning:

- Students spend time exploring problems in depth.
- They find more than one solution to many of the problems they work on.
- They invent their own strategies and approaches, rather than rely on memorized procedures.
- They choose from a variety of concrete materials and appropriate technology, including calculators, as a natural part of their everyday mathematical work.
- They express their mathematical thinking through drawing, writing, and talking.
- They work in a variety of groupings—as a whole class, individually, in pairs, and in small groups.
- They move around the classroom as they explore the mathematics in their environment and talk with their peers.

While reading and other language activities are typically given a great deal of time and emphasis in elementary classrooms, mathematics often does not get the time it needs. If students are to experience mathematics in depth, they must have enough time to become engaged in real mathematical problems. We believe that a minimum of 5 hours of mathematics classroom time a week—about an hour a day—is critical at the elementary level. The scope and pacing of the *Investigations* curriculum are based on that belief.

We explain more about the pedagogy and principles that underlie these investigations in Teacher Notes throughout the units. For correlations of the curriculum to the NCTM Standards and further help in using this research-based program for teaching mathematics, see the following books, available from Dale Seymour Publications:

- *Implementing the* Investigations in Number, Data, and Space® *Curriculum*
- *Beyond Arithmetic: Changing Mathematics in the Elementary Classroom* by Jan Mokros, Susan Jo Russell, and Karen Economopoulos

HOW TO USE THIS BOOK

This book is one of the curriculum units for *Investigations in Number, Data, and Space*. In addition to providing part of a complete mathematics curriculum for your students, this unit offers information to support your own professional development. You, the teacher, are the person who will make this curriculum come alive in the classroom; the book for each unit is your main support system.

Although the curriculum does not include student instructional texts, reproducible sheets for student work are provided with the units and, in some cases, are also available as Student Activity Booklets. In these investigations, students work actively with objects and experiences in their own environment, including manipulative materials and technology, rather than with a workbook.

Ultimately, every teacher will use these investigations in ways that make sense for his or her particular style, the particular group of students, and the constraints and supports of a particular school environment. Each unit offers information and guidance drawn from our collaborations with many teachers and students over many years. Our goal is to help you, a professional educator, give all your students access to mathematical power.

Investigation Format

The opening two pages of each investigation help you get ready for the work that follows.

- **Focus Time** This gives a synopsis of the activities used to introduce the important mathematical ideas for the investigation.
- **Choice Time** This lists the activities, new and recurring, that support the Focus Time work.
- **Mathematical Emphasis** This highlights the most important ideas and processes students will encounter in this investigation.
- **Teacher Support** This indicates the Teacher Notes and Dialogue Boxes included to help you understand what's going on mathematically in your classroom.
- **What to Plan Ahead of Time** These lists alert you to materials to gather, sheets to duplicate, and other things you need to do before starting the investigation. Full details of materials and preparation are included with each activity.

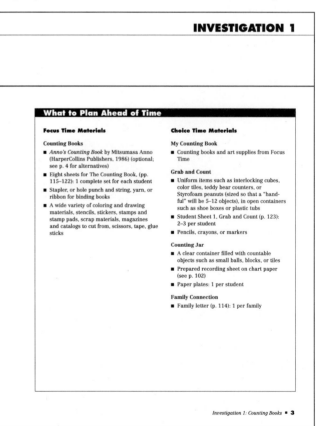

Always read through an entire investigation before you begin, in order to understand the overall flow and sequence of the activities.

Focus Time In this whole-group meeting, you introduce one or more activities that embody the important mathematical ideas underlying the investigation. The group then may break up into individuals or pairs for further work on the same activity. Many Focus Time activities culminate with a brief sharing time or discussion as a way of acknowledging students' work and highlighting the mathematical ideas. Focus Time varies in length. Sometimes it is short and can be completed in a single group meeting or a single work period; other times it may stretch over two or three sessions.

Choice Time Each Focus Time is followed by Choice Time, which offers a series of supporting activities to be done simultaneously by individuals, pairs, or small groups. You introduce these related tasks over a period of several days. During Choice Time, students work independently, at their own pace, choosing the activities they prefer and often returning many times to their favorites. Many kindergarten classrooms have an activity time built into their daily schedule, and Choice Time activities can easily be incorporated.

Together, the Focus Time and Choice Time activities offer a balanced kindergarten curriculum.

Classroom Routines The kindergarten day is filled with opportunities to work with mathematics. Routines such as taking attendance, asking about snack preferences, and discussing the calendar offer regular, ongoing practice in counting, collecting and organizing data, and understanding time.

Four specific routines—Attendance, Counting Jar, Calendar, and Today's Question—are formally introduced in the unit *Mathematical Thinking in Kindergarten*. Another routine, Patterns on the Pocket Chart, is introduced in the unit *Pattern Trains and Hopscotch Paths*. Descriptions of these routines can also be found in an appendix for each unit, and reminders of their ongoing use appear in the Unit Overview charts.

The Linguistically Diverse Classroom Each unit includes an appendix with Tips for the Linguistically Diverse Classroom to help teachers support students at varying levels of English proficiency. While more specific tips appear within the units at grades 1–5, often in relation to written work, general tips on oral discussions and observing the students are more appropriate for kindergarten.

Also included are suggestions for vocabulary work to help ensure that students' linguistic difficulties do not interfere with their comprehension of math concepts. The Preview for the Linguistically Diverse Classroom lists key words in the unit that are generally known to English-speaking kindergartners. Activities to help familiarize other students with these words are found in the appendix, Vocabulary Support for Second-Language Learners. In addition, ideas for making connections to students' languages and cultures, included on the Preview page, help the class explore the unit's concepts from a multicultural perspective.

Materials

A complete list of the materials needed for teaching this unit follows the Unit Overview. These materials are available in *Investigations* kits or can be purchased from school supply dealers.

Classroom Materials In an active kindergarten mathematics classroom, certain basic materials should be available at all times, including interlocking cubes, a variety of things to count with, and writing and drawing materials. Some activities in this curriculum require scissors and glue sticks or tape; dot stickers and large paper are also useful. So that students can independently get what they need at any time, they should know where the materials are kept, how they are stored, and how they are to be returned to the storage area.

Children's Literature Each unit offers a list of children's literature that can be used to support the mathematical ideas in the unit. Sometimes an activity incorporates a specific children's book, with suggestions for substitutions where practical. While such activities can be adapted and taught without the book, the literature offers a rich introduction and should be used whenever possible. If you can get the titles in Big Book format, these are ideal for kindergarten.

Blackline Masters Student recording sheets and other teaching tools for both class and homework are provided as reproducible blackline masters at the end of each unit. When student sheets are designated for kindergarten homework, they usually repeat an activity from class, such as playing a game, as a way of involving and informing family members. Occasionally a homework sheet may ask students to collect data or materials for a class project or in preparation for upcoming activities.

Student Activity Booklets For the two kindergarten number units, the blackline masters are also available as Student Activity Booklets, designed to free you from extensive copying. The other kindergarten units require minimal copying.

Family Letter A letter that you can send home to students' families is included with the blackline masters for each unit. Families need to be informed about the mathematics work in your classroom; they should be encouraged to participate in and support their children's work. A reminder to send home the letter for each unit appears in one of the early investigations. These letters are also available separately in Spanish, Vietnamese, Cantonese, Hmong, and Cambodian.

Investigations **at Home** To further involve families in the kindergarten program, you can offer them the *Investigations* at Home booklet, which describes the kindergarten units, explains the mathematics work children do in kindergarten, and offers activities families can do with their children at home.

Adapting *Investigations* to Your Classroom

Kindergarten programs vary greatly in the amount of time each day that students attend. We recommend that kindergarten teachers devote from 30 to 45 minutes daily to work in mathematics, but we recognize that this can be challenging in a half-day program. The kindergarten level of *Investigations* is intentionally flexible so that teachers can adapt the curriculum to their particular setup.

Kindergartens participating in the *Investigations* field test included full-day programs, half-day programs of approximately 3 hours, and half-day programs that add one or two full days to the kindergarten week at some point in the school year. Despite the wide range of program structures, classrooms generally fell into one of two groups: those that offered a separate math time daily (Math Workshop or Math Time), and those that included one or two mathematics activities during a general Activity Time or Station Time.

Math Workshop Teachers using a Math Workshop approach set aside 30 to 45 minutes each day for doing mathematics. In addition, they usually also have a more general activity time in their daily schedule. On some days, Math Workshop might be devoted to the Focus Time activities, with the whole class gathered together. On other days, students might work in small groups and choose from three or four Choice Time activities.

Math as Part of Activity Time Teachers with less time in their day may offer students one or two math activities, along with activities from other areas of the curriculum, during their Activity Time or Station Time. For example, on a particular day, students might be able to choose among a science activity, block building, an art project, dramatic play, books, puzzles, and a math activity. New activities are introduced during a whole-class meeting. With the *Investigations* curriculum, teachers who use this approach have found that it is important to designate at least one longer block of time (30 to 45 minutes) each week for mathematics. During this time, students engage in Focus Time activities and have a chance to share their work and discuss mathematical ideas. The suggested Choice Time activities are then presented as part of the general activity time. Following this model, work on a curriculum unit will naturally stretch over a longer period.

Planning Your Curriculum The amount of time scheduled for mathematics work will determine how much of the kindergarten *Investigations* curriculum a teacher is able to cover in the school year. You may have to make some choices as you adapt the units to your particular schedule. What is most important is finding a way to involve students in mathematics every day of the school year.

Each unit will be handled somewhat differently by every teacher. You need to be active in determining an appropriate pace and the best transition points for your class. As you read an investigation, make some preliminary decisions about how many days you will need to present the activities, based on what you know about your students and about your schedule. You may need to modify your initial plans as you proceed, and you may want to make notes in the margins of the pages as reminders for the next time you use the unit.

Help for You, the Teacher

Because we believe strongly that a new curriculum must help teachers think in new ways about mathematics and about their students' mathematical thinking processes, we have included a great deal of material to help you learn more about both.

About the Mathematics in This Unit This introductory section summarizes the essential information about the mathematics you will be teaching. It describes the unit's central mathematical ideas and the ways students will encounter them through the unit's activities.

Teacher Notes These reference notes provide practical information about the mathematics you are teaching and about our experience with how students learn. Many of the notes were written in response to actual questions from teachers or to discuss important things we saw happening in the field-test classrooms. Some teachers like to read them all before starting the unit, then review them as they come up in particular investigations.

In the kindergarten units, Teacher Notes headed "From the Classroom" contain anecdotal reflections of teachers. Some focus on classroom management issues, while others are observations of students at work. These notes offer another perspective on how an activity might unfold or how kindergarten students might become engaged with a particular material or activity.

A few Teacher Notes touch on fundamental principles of using *Investigations* and focus on the pedagogy of the kindergarten classroom:

- About Choice Time
- Materials as Tools for Learning
- Encouraging Students to Think, Reason, and Share Ideas
- Games: The Importance of Playing More Than Once

After their initial appearance, these are repeated in the back of each unit. Reviewing these notes periodically can help you reflect on important aspects of the *Investigations* curriculum.

Dialogue Boxes Sample dialogues demonstrate how students typically express their mathematical ideas, what issues and confusions arise in their thinking, and how some teachers have guided class discussions.

Many of these dialogues are word-for-word transcriptions of recorded class discussions. They are not always easy reading; sometimes it may take some effort to unravel what the students are trying to say. But this is the value of these dialogues; they offer good clues to how your students may develop and express their approaches and strategies, helping you prepare for your own class discussions.

Where to Start You may not have time to read everything the first time you use this unit. As a first-time user, you will likely focus on understanding the activities and working them out with your students. You will also want to read the few sections listed in the Contents under the heading Where to Start.

TECHNOLOGY IN THE CURRICULUM

The *Investigations* curriculum incorporates the use of two forms of technology in the classroom: calculators and computers. Calculators are assumed to be standard classroom materials, available for student use in any unit. Computers are explicitly linked to one or more units at each grade level; they are used with the unit on 2-D geometry at each grade, as well as with some of the units on measuring, data, and changes.

Using Calculators

In this curriculum, calculators are considered tools for doing mathematics, similar to pattern blocks or interlocking cubes. Just as with other tools, students must learn both *how* to use calculators correctly and *when* they are appropriate to use. This knowledge is crucial for daily life, as calculators are now a standard way of handling numerical operations, both at work and at home. Calculators are formally introduced in the grade 1 curriculum, but if available, can be introduced to kindergartners informally.

Using a calculator correctly is not a simple task; it depends on a good knowledge of the four operations and of the number system, so that students can select suitable calculations and also determine what a reasonable result would be. These skills are the basis of any work with numbers, whether or not a calculator is involved.

Unfortunately, calculators are often seen as tools to check computations with, as if other methods are somehow more fallible. Students need to understand that any computational method can be used to check any other; it's just as easy to make a mistake on the calculator as it is to make a mistake on paper or with mental arithmetic. Throughout this curriculum, we encourage students to solve computation problems in more than one way in order to double-check their accuracy. We present mental arithmetic, paper-and-pencil computation, and calculators as three possible approaches.

In this curriculum we also recognize that, despite their importance, calculators are not always appropriate in mathematics instruction. Like any tools, calculators are useful for some tasks but not for others. You will need to make decisions about when to allow students access to calculators and when to ask that they solve problems without them, so that they can concentrate on other tools and skills. At times when calculators are or are not appropriate for a particular activity, we make specific recommendations. Help your students develop their own sense of which problems they can tackle with their own reasoning and which ones might be better solved with a combination of their own reasoning and the calculator.

Managing calculators in your classroom so that they are a tool, and not a distraction, requires some planning. When calculators are first introduced, students often want to use them for everything, even problems that can be solved quite simply by other methods. However, once the novelty wears off, students are just as interested in developing their own strategies, especially when these strategies are emphasized and valued in the classroom. Over time, students will come to recognize the ease and value of solving problems mentally, with paper and pencil, or with manipulatives, while also understanding the power of the calculator to facilitate work with larger numbers.

Experience shows that if calculators are available only occasionally, students become excited and distracted when permitted to use them. They focus on the tool rather than on the mathematics. In order to learn when calculators are appropriate and when they are not, students must have easy access to them and use them routinely in their work.

If you have a calculator for each student, and if you think your students can accept the responsibility, you might allow them to keep their calculators with the rest of their individual materials, at least for the first few weeks of school. Alternatively, you might store them in boxes on a shelf, number each calculator, and assign a corresponding number to each student. This system can give students a sense of ownership while also helping you keep track of the calculators.

Using Computers

Students can use computers to approach and visualize mathematical situations in new ways. The computer allows students to construct and manipulate geometric shapes, see objects move according to rules they specify, and turn, flip, and repeat a pattern.

This curriculum calls for computers in units where they are a particularly effective tool for learning mathematics content. One unit on 2-D geometry at each of the grades 3–5 includes a core of activities that rely on access to computers, either in the classroom or in a lab. Other units on geometry, measurement, data, and changes include computer activities, but can be taught without them. In these units, however, students' experience is greatly enhanced by computer use.

The following list outlines the recommended use of computers in this curriculum:

Kindergarten
Unit: *Making Shapes and Building Blocks* (Exploring Geometry)
Software: *Shapes*
Source: provided with the unit

Grade 1
Unit: *Survey Questions and Secret Rules* (Collecting and Sorting Data)
Software: *Tabletop, Jr.*
Source: Broderbund

Unit: *Quilt Squares and Block Towns* (2-D and 3-D Geometry)
Software: *Shapes*
Source: provided with the unit

Grade 2
Unit: *Mathematical Thinking at Grade 2* (Introduction)
Software: *Shapes*
Source: provided with the unit

Unit: *Shapes, Halves, and Symmetry* (Geometry and Fractions)
Software: *Shapes*
Source: provided with the unit

Unit: *How Long? How Far?* (Measuring)
Software: *Geo-Logo*
Source: provided with the unit

Grade 3
Unit: *Flips, Turns, and Area* (2-D Geometry)
Software: *Tumbling Tetrominoes*
Source: provided with the unit

Unit: *Turtle Paths* (2-D Geometry)
Software: *Geo-Logo*
Source: provided with the unit

Grade 4
Unit: *Sunken Ships and Grid Patterns* (2-D Geometry)
Software: *Geo-Logo*
Source: provided with the unit

Grade 5
Unit: *Picturing Polygons* (2-D Geometry)
Software: *Geo-Logo*
Source: provided with the unit

Unit: *Patterns of Change* (Tables and Graphs)
Software: *Trips*
Source: provided with the unit

Unit: *Data: Kids, Cats, and Ads* (Statistics)
Software: *Tabletop, Sr.*
Source: Broderbund

The software provided with the *Investigations* units uses the power of the computer to help students explore mathematical ideas and relationships that cannot be explored in the same way with physical materials. With the *Shapes* (grades K–2) and *Tumbling Tetrominoes* (grade 3) software, students explore symmetry, pattern, rotation and reflection, area, and characteristics of 2-D shapes. With the *Geo-Logo* software (grades 2–5), students investigate rotations and reflections, coordinate geometry, the properties of 2-D shapes, and angles. The *Trips* software (grade 5) is a mathematical exploration of motion in which students run experiments and interpret data presented in graphs and tables.

We suggest that students work in pairs on the computer; this not only maximizes computer resources but also encourages students to consult, monitor, and teach one another. However, asking more than two students to work at the same computer is less effective. Managing access to computers is an issue for every classroom. The curriculum gives you explicit support for setting up a system. The units are structured on the assumption that you have enough computers for half your students to work on the machines in pairs at one time. If you do not have access to that many computers, suggestions are made for structuring class time to use the unit with fewer than five.

ABOUT ASSESSMENT

Assessment plays a critical role in teaching and learning, and it is an integral part of the *Investigations* curriculum. For a teacher using these units, assessment is an ongoing process. You observe students' discussions and explanations of their ideas and strategies on a daily basis and examine their work as it evolves. While students are busy working with materials, playing mathematical games, sharing ideas with partners, and working on projects, you have many opportunities to observe their mathematical thinking. What you learn through observation guides your decisions about how to proceed, both with the curriculum and with individual students.

Our experiences with young children suggest that they know, can explain, and can demonstrate with materials a lot more than they can represent on paper. This is one reason why it is so important to engage children in conversation, helping them explain their thinking about a problem they are solving. It is also why, in kindergarten, assessment is based exclusively on a teacher's observations of students as they work.

The way you observe students will vary throughout the year. At times you may be interested in particular strategies that students are developing to solve problems. Other times, you might want to observe how students use or do not use materials for solving problems. You may want to focus on how students interact when working in pairs or groups. You may be interested in noting the strategy that a student uses when playing a game during Choice Time. Or you may take note of student ideas and thinking during class discussions.

Assessment Tools in the Unit

Virtually every activity in the kindergarten units of the *Investigations* curriculum includes a section called Observing the Students. This section is a teacher's primary assessment tool. It offers guidelines on what to look for as students encounter the mathematics of the activity. It may suggest questions you can ask to uncover student thinking or to stimulate further investigation. When useful, a range of potential responses or examples of typical student approaches are given, along with ways to adapt the activity for students in need of more or less challenge.

Supplementing this main assessment tool in each unit are the Teacher Notes and Dialogue Boxes that contain examples of student work, teacher observations, and student conversations from real kindergarten classrooms. These resources can help you interpret experiences from your own classroom as you progress through a unit.

Documentation of Student Growth

You will probably need to develop some sort of system to record and keep track of your observations. A single observation is like a snapshot of a student's experience with a particular activity, but when considered over time, a collection of these snapshots provides an informative and detailed picture of a student. Such observations are useful in documenting and assessing student's growth, as well as in planning curriculum.

Observation Notes A few ideas that teachers have found successful for record keeping are suggested here. The most important consideration is finding a system that really works for you. All too often, keeping observation notes on a class of 20–30 students is overwhelming and time-consuming. Your goal is to find a system that is neither.

Some teachers find that a class list of names is convenient for jotting down their observations. Since the space is limited, it is not possible to write lengthy notes; however, over time, these short observations provide important information.

Other teachers keep a card file or a loose-leaf notebook with a page for each student. When something about a student's thinking strikes them as important, they jot down brief notes and the date.

Some teachers use self-sticking address labels, kept on clipboards around the classroom. After taking notes on individual students, they simply peel off each label and stick it into the appropriate student file or notebook page.

You may find that writing notes at the end of each week works well for you. For some teachers, this process helps them reflect on individual students, on the curriculum, and on the class as a whole. Planning for the next weeks' activities often grows out of these weekly reflections.

Student Portfolios Collecting samples of student work from each unit in a portfolio is another way to document a student's experience that supports your observation notes. In kindergarten, samples of student work may include constructions, patterns, or designs that students have recorded, score sheets from games they have played, and early attempts to record their problem-solving strategies on paper, using pictures, numbers, or words.

The ability to record and represent one's ideas and strategies on paper develops over time. Not all 5- and 6-year-olds will be ready for this. Even when students are ready, what they record will have meaning for them only in the moment—as they work on the activity and make their representation. You can augment this by taking dictation of a student's idea or strategy. This not only helps both you and the student recall the idea, but also gives students a model of how their ideas could be recorded on paper.

Over the school year, student work samples combined with anecdotal observations are valuable resources when you are preparing for family conferences or writing student reports. They help you communicate student growth and progress, both to families and to the students' subsequent teachers.

Assessment Overview

There are two places to turn for a preview of the assessment information in each kindergarten *Investigations* unit. The Assessment Resources column in the Unit Overview chart locates the Observing the Students section for each activity, plus any Teacher Notes and Dialogue Boxes that explain what to look for and what types of responses you might see in your classroom. Additionally, the section called About the Assessment in This Unit gives you a detailed list of questions, keyed to the mathematical emphases for each investigation, to help you observe and assess student growth. This section also includes suggestions for choosing student work to save from the unit.

These examples illustrate record keeping systems used by two different teachers for the kindergarten unit *Collecting, Counting, and Measuring*, one using the class list and the other using individual note cards to record student progress.

Emma Ruiz

3/19 Counting Jar: counts 9 balls accurately and makes another set of 9 cubes

3/24 Today's Question: compares data, "13 is 4 more than 9 because the 13 tower is 4 names taller."

4/1 Draws counting book pictures for 1–6, then adds pgs 7, 8, 9, 10, 11 on her own

---

Unit: Collecting, Counting, and Measuring
Activity: Inventory Bags
Date: 10/12 and 10/13

Alexa • counting sequence to 50↑ • counts 1:1 up to 12 • counts 4 bags accurately	**Luke** • counts to 30, misses 19, 20 and 29, 30 • counts by moving objects; 1:1 to 10 objects • draws circles for buttons
Ayesha • works with Oscar • counts to 15 accurately – trouble beyond 15 but Oscar helps ★meet to check • 1:1 to 8 objects? counting	**Maddy** • difficult to tell how much M. counted herself + how much was done by partner. Work w/ her to see.
Brendan absent 10/12, 10/13	**Miyuki** • counts aloud beyond 30 but leaves out 14 • counts 1:1 up to 10 but doesn't organize objects
Carlo • counts objects with difficulty. • remove items from bag so he works with 10 • says numbers to 10, counts objects to 6	**Oscar** • works with Ayesha • counts rotely to 20, maybe higher • double-checks his count every time – is accurate
Charlotte • completed inventory task easily without help • counts accurately up to 20 objects • represents with numbers	**Ravi** • worked w/ his aide to complete task • counts 1:1 to 5 objects • difficulty representing quantity w/ pictures
Felipe • worked well with Tarik • counted ~~_____~~ bag – 21 in all	**Renata**

UNIT OVERVIEW

Collecting, Counting, and Measuring

Content of This Unit In this unit students explore numbers and number relationships through a variety of counting experiences. They count sets of classroom materials such as interlocking cubes, pencils, color tiles, and art supplies. They make counting books, and count and compare the number of letters in their names. They play mathematical games and solve mathematical problems that involve counting and accumulating amounts. With repeated opportunities to count in real ways, students build their knowledge of the counting sequence and of the quantities those numbers represent. Throughout the unit students begin to explore ways to use pictures, numerals, objects, and words to represent the quantities they count. In addition to their work with counting, students explore the concept of measurement as they directly compare the lengths of objects to find out which is longer.

Connections with Other Units If you are doing the full-year *Investigations* curriculum in the suggested sequence for kindergarten, this is the third of six units. The work extends the counting activities introduced in *Mathematical Thinking in Kindergarten* and provides the foundation for future work with counting and measurement in *Counting Ourselves and Others* and *How Many in All?*

Investigations Curriculum ■ Suggested Kindergarten Sequence

Mathematical Thinking in Kindergarten (Introduction)

Pattern Trains and Hopscotch Paths (Exploring Pattern)

▶ *Collecting, Counting, and Measuring* (Developing Number Sense)

Counting Ourselves and Others (Exploring Data)

Making Shapes and Building Blocks (Exploring Geometry)

How Many in All? (Counting and the Number System)

Investigation 1 ■ Counting Books

Class Sessions	Activities	Pacing
FOCUS TIME (p. 4) Counting Books	*Anno's Counting Book* Make-Your-Own Counting Books Extension: Seasons Extension: Time Extension: Forming a Community Homework: Family Connection	1–2 sessions
CHOICE TIME (p. 10)	My Counting Book Grab and Count Counting Jar	2–3 sessions
Classroom Routines	Attendance and Calendar (daily) Counting Jar, Today's Question, and Patterns on the Pocket Chart (weekly or as appropriate)	

Mathematical Emphasis

- Thinking about what, when, and why people count
- Recognizing numerals and number names
- Connecting numerals to the quantities they represent
- Creating a set of a given size
- Developing strategies for counting and keeping track of quantities
- Representing quantities with pictures, numerals, or words

Assessment Resources

Observing the Students:

- Make-Your-Own Counting Books (p. 8)
- My Counting Book (p. 11)
- Grab and Count (p. 13)
- Counting Jar (p. 15)

Teacher Note: Counting Is More Than 1, 2, 3 (p. 16)

Teacher Note: Students' Counting Books (p. 19)

Materials

Anno's Counting Book or a similar book

Coloring and drawing materials, stencils, stickers, rubber stamps and stamp pads

Stapler or hole-punch and string

Open containers of cubes, tiles, teddy bear counters

Clear container (Counting Jar)

Various countable objects

Paper plates

Chart paper

Student Sheet 1

Teaching resource sheets

Counting Jar

Investigation 2 ■ Taking Inventory

Class Sessions	Activities	Pacing
FOCUS TIME (p. 24) Taking Inventory	Counting What's in the Bag What Did You Find Out?	1–2 sessions
CHOICE TIME (p. 30)	Inventory Bags My Counting Book Grab and Count Counting Jar	2–3 sessions
Classroom Routines	Attendance and Calendar (daily) Counting Jar, Today's Question, and Patterns on the Pocket Chart (weekly or as appropriate)	

Mathematical Emphasis

- Connecting numerals to the quantities they represent
- Developing strategies for counting and keeping track of quantities
- Creating a set of a given size
- Representing quantities with pictures, numerals, or words

Assessment Resources

Observing the Students:

- Inventory Bags (pp. 26 and 31)

Teacher Note: Observing Students As They Count (p. 32)

Teacher Note: Students' Difficulties with Counting (p. 34)

Materials

Paper bags
Various countable items
Students' Counting Books
Coloring and drawing materials, stencils, stickers, rubber stamps and stamp pads
Open containers of cubes, tiles, teddy bear counters
Counting Jar
Paper plates
Chart paper
Student Sheets 1–2
Teaching resource sheets

Teddy bear counters

Investigation 3 ■ Comparing Towers

Class Sessions	Activities	Pacing
FOCUS TIME (p. 38) Measurement Towers	Using Towers to Compare and Measure How Did You Compare? Extension: A Longer/Shorter Hunt Homework: Compare at Home	2 sessions
CHOICE TIME (p. 42)	Measuring Table Grab and Count: Which Has More? Compare	3–4 sessions
Classroom Routines	Attendance and Calendar (daily) Counting Jar, Today's Question, and Patterns on the Pocket Chart (weekly or as appropriate)	

Mathematical Emphasis

- Developing and using language to describe and compare lengths
- Measuring by direct comparison
- Developing strategies for counting and keeping track of quantities
- Representing quantities with objects, pictures, numerals, or words
- Comparing two quantities to find which is more
- Recognizing numerals and connecting them with the quantities they represent

Assessment Resources

Observing the Students:

- Measuring Table (p. 43)
- Grab and Count: Which Has More? (p. 45)
- Compare (p. 47)

Teacher Note: Learning About Length (p. 50)

Materials

Interlocking cubes

Crayons, colored pencils, markers

Collection of objects to measure

Number Cards

Teaching resource sheets

Number Cards

Unit Overview ■ I-13

Investigation 4 ▪ Counting and Comparing

Class Sessions	Activities	Pacing
FOCUS TIME (p. 54) Letters in Our Names	*Chrysanthemum* How Many Letters in Your Name? Making Name Towers Extension: First, Middle, and Last Names Extension: How Many Letters in Our Names? Homework: Names at Home	1–2 sessions
CHOICE TIME (p. 60)	Comparing Names Grab and Count: Compare Collect 10 Together	3–4 sessions
Classroom Routines	Attendance and Calendar (daily) Counting Jar, Today's Question, and Patterns on the Pocket Chart (weekly or as appropriate)	

Mathematical Emphasis

- Counting groups of objects
- Creating a set of a given size
- Comparing quantities to determine which is more
- Using terms to describe and compare amounts
- Keeping track of the size of a growing collection
- Finding the total of two single-digit numbers
- Recording mathematical work

Assessment Resources

Observing the Students:

- Making Name Towers (p. 57)
- Comparing Names (p. 61)
- Grab and Count: Compare (p. 63)
- Collect 10 Together (p. 65)

Materials

Name cards (or index cards)
Chrysanthemum (optional)
Interlocking cubes
Crayons, colored pencils, markers
Dot stickers
Chart paper (optional)
Dot cubes with 1–3 dots
Counters
Student Sheet 3
Teaching resource sheet

Interlocking cubes

I-14 ▪ *Collecting, Counting, and Measuring*

Investigation 5 ■ Least to Most

Class Sessions	Activities	Pacing
FOCUS TIME (p. 68) Least to Most	Four Handfuls: Least to Most Talking About Least to Most Homework: Collect 10 Together at Home	2 sessions
CHOICE TIME (p. 72)	Grab and Count: Least to Most Racing Bears Collect 10 Together	2–3 sessions
Classroom Routines	Attendance and Calendar (daily) Counting Jar, Today's Question, and Patterns on the Pocket Chart (weekly or as appropriate)	

Mathematical Emphasis

- Counting sets of objects
- Comparing quantities to determine which is more
- Using appropriate language to describe and compare amounts
- Ordering quantities from least to most or most to least
- Representing mathematical work

Assessment Resources

Observing the Students:

- Grab and Count: Least to Most (p. 73)
- Racing Bears (p. 75)

Dialogue Box: Comparing and Ordering Towers (p. 76)

Materials

Interlocking cubes
Crayons, colored pencils, markers
Glue sticks or tape
Unlined paper
Teddy bear counters
Other counters
Dot cubes (1–3 and 1–6 dots)
Teaching resource sheets

Racing Bears gameboard

Investigation 6 ■ Arrangements of Six

Class Sessions	Activities	Pacing
FOCUS TIME (p. 80) Six Tiles	Arranging Six Tiles My Six Tiles Sharing Our Arrangements Extension: Sorting Arrangements of Six Extension: 3-D Arrangements of Six	1–2 sessions
CHOICE TIME (p. 86)	Books of Six Collect 10 Together Grab and Count: Least to Most Racing Bears	3–4 sessions
Classroom Routines	Attendance and Calendar (daily) Counting Jar, Today's Question, and Patterns on the Pocket Chart (weekly or as appropriate)	

Mathematical Emphasis

- Finding ways to visualize and arrange a set of six objects
- Solving a problem with many possible solutions
- Developing strategies for counting and keeping track of quantities through about 12
- Representing quantities with objects, pictures, or numerals
- Comparing quantities to determine which is more
- Using language to describe and compare amounts
- Ordering from least to most

Assessment Resources

Observing the Students:

- Arranging Six Tiles (p. 81)
- My Six Tiles (p. 83)
- Books of Six (p. 87)

Teacher Note: "I Had One Too Much" (p. 89)

Materials

Color tiles
Construction paper squares
Glue sticks
Interlocking cubes
Teddy bear counters
Other counters
Dot cubes (1–3 and 1–6 dots)
Stapler or hole-punch and string
Unlined paper
Teaching resource sheets

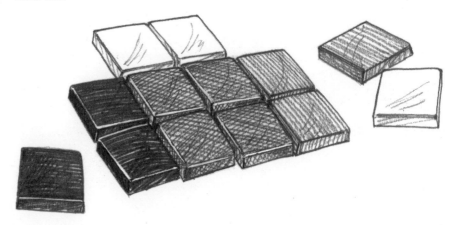

Collecting, Counting, and Measuring

MATERIALS LIST

Following are the basic materials needed for the activities in this unit. Many items can be purchased from the publisher, either individually or in the Teacher Resource Package and the Student Materials Kit for kindergarten. Detailed information is available on the *Investigations* order form. To obtain this form, call toll-free 1-800-872-1100 and ask for a Dale Seymour customer service representative.

Snap™ Cubes (interlocking cubes that connect on all sides): class set, or 1 tub of 100 per 4–6 students

Color tiles: 2 sets of 400 per class

Teddy bear counters: 1–2 sets

Collections of countable materials, such as pattern blocks, Geoblocks, Styrofoam peanuts, pencils, erasers, keys, buttons, shells, pebbles, lids

Blank 1-inch cubes, with stick-on labels for making large dot cubes

Counters, such as cubes, buttons, or bread tabs: 15–20 per pair

Primary Number Cards: 1 deck per pair (manufactured, or use blackline masters to make your own)

Clear plastic container to use as a Counting Jar

Anno's Counting Book by Mitsumasa Anno (optional)

Chrysanthemum by Kevin Henkes (optional)

Paper plates: 1 per student

Small paper bags

Index cards (or other student name cards)

Colored pencils, markers, crayons

Assorted art supplies, such as stencils, stickers (including dot stickers), rubber stamps and stamp pads, magazines and catalogs

Glue sticks or tape

Scissors

Colored construction paper

Chart paper

Unlined paper

Stapler or hole-punch and string or ribbon, for binding student books

The following materials are provided at the end of this unit as blackline masters.

Family Letter (p. 114)

Student Sheets 1–3 (p. 123)

Teaching Resources:

   The Counting Book (p. 115)

   How to Play Compare (p. 125)

   Number Cards (p. 126)

   How to Play Collect 10 Together (p. 131)

   Paper Towers for Collect 10 (p. 132)

   Racing Bears Gameboard (p. 133)

   One-Inch Grid Paper (p. 134)

   Cube Strips for Grab and Count (p. 135)

Related Children's Literature

Counting Books*

Anno, Mitsumasa. *Anno's Counting Book.* New York: HarperCollins Publishers, 1986.

Crews, Donald. *Ten Black Dots.* New York: Greenwillow Books, 1986.

Falwell, Cathryn. *Feast for Ten.* New York: Clarion Books, 1993.

Hoban, Tana. *Count and See.* New York: Macmillan, 1972.

Sloat, Teri. *From One to One Hundred.* New York: A Puffin Unicorn Book, 1991.

Books About Long Names

Henkes, Kevin. *Chrysanthemum.* New York: Greenwillow Books, 1991.

Mosel, Arlene. *Tikki Tikki Tembo.* New York: Greenwillow Books, 1968.

Zelinsky, Paul O. *Rumpelstiltskin.* New York: Dutton Children's Press, a division of Penguin Books, 1986.

* See also p. I-22 for suggested multicultural counting books.

ABOUT THE MATHEMATICS IN THIS UNIT

Collecting, Counting, and Measuring gives students many meaningful opportunities to develop their sense of numbers and quantities, to count, and to measure objects by comparing them directly.

Number Sense Throughout this unit, students are developing number sense by counting and comparing quantities. Just as common sense grows from our experience with the world and how it works, *number sense* grows from experience with numbers and how they work. Students in the same kindergarten class can vary considerably in age and in their previous experience with numbers. While many students entering kindergarten may know the oral counting sequence up to 5, 10, or even higher, they will vary tremendously in their ability to accurately count out a set of objects and in their sense of the size of quantities. Many are just beginning to explore the ways in which numbers give information about quantities of real things.

Through the counting activities in this unit, students learn and practice the rote counting sequence (the number names and their order) and learn to recognize the written numerals. They work to establish a strong connection between spoken and written numbers and the quantities they represent, so that when they hear or see "3," they know that it represents a group of three things. They also explore another important idea: that three is *always* three, whether it is three objects pushed or linked together, or three objects spread apart in a line or some other formation.

Counting Counting involves more than knowing the number names, their sequence, and how to write numerals. In order to count successfully, students must remember the counting sequence, assign one number to each object as they count, and at the same time keep track of what they are counting.

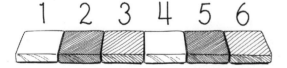

Students develop their understanding of quantity through repeated experiences with organizing and counting sets of objects. In kindergarten, most of the activities can be adjusted to provide an appropriate level of challenge for individual students. Early in the year, some students will need repeated experiences with quantities up to 5, while others will be able to work with larger collections.

An important part of counting a set of objects accurately is saying one number name for each object that is counted. When students first begin to count objects, they often do not connect the numbers they say to the objects they are counting. You might see students who point randomly at the objects, or even in the air, but do not seem to connect the number of objects in the pile with the last number in their count. Students learn about one-to-one correspondence through many opportunities to count sets of objects and to watch others as they count.

Another important piece of counting accurately is being able to keep track of the objects in order to know what has been counted and what still remains to be counted. As students first begin to count sets of objects, they often do not organize them. As a result, they will count some objects more than once and skip others altogether. Students develop strategies for keeping track of a count as they realize the need. In this unit, students use a variety of manipulative materials that help them see the need for and develop such strategies.

Measuring by Direct Comparison When students count a tower of cubes to find out "how many," a natural question often arises: Which tower has more—or which is longer—and how can we find out? Comparing is a natural way for students to approach measuring. Young children often spontaneously try to see who or what is bigger, taller, longer, or smaller. This kind of qualitative comparison is essential as a foundation for developing ideas about length.

In this unit, students directly compare the length of two objects, such as two different towers of cubes, or a book and a 10-cube tower. In these comparisons, students are working on a key idea about measuring—a single characteristic of an object may not tell you everything about its measure. Comparing end points of two objects might indicate which is longer, but only if the beginning points are aligned. Kindergarten students are just developing ideas about what they need to pay attention to in order to decide which of two things is longer.

Developing language about measuring and comparing is another emphasis of this unit. Students need to become comfortable using their own language to describe their measuring activities and to hear and use words that describe length—*long, short, wide, tall, high*—and the comparative forms of the words—*longer, wider, taller,* and so forth. In discussions, you'll discover that students develop and use a wide range of strategies for comparing the length of two objects. Keep in mind that trial and error is age-appropriate and may be a predominant strategy for many children.

Mathematical Emphasis At the beginning of each investigation, the mathematical emphasis section tells you what is most important for students to learn about during that investigation. Many of these understandings and processes are difficult and complex. Students gradually learn more and more about each idea over many years of schooling. Individual students will begin and end the unit with different levels of knowledge and skill, but all will work toward developing good number sense through counting, comparing, and measuring.

ABOUT THE ASSESSMENT IN THIS UNIT

Throughout the *Investigations* curriculum, there are many opportunities for ongoing daily assessment as you observe, listen to, and interact with students at work. You can use almost any activity in this unit to assess your students' needs and strengths. Listed below are questions to help you focus your observations in each investigation. You may want to keep track of your observations for each student to help you plan your curriculum and monitor students' growth. Suggestions for documenting student growth can be found in the section About Assessment (p. I-8).

Investigation 1: Counting Books

- Are students able to name instances when people count? Can they tell you what is being counted and why?
- How do students count orally? Do they know the sequence of number names? Do they forget any of them? Which ones? Do they skip certain numbers or mix the order of the names? Which ones? Do they self-correct? Do they hear mistakes in the counting sequence if someone else makes them?
- Do students recognize the numerals? Up to what number? Do they connect the number names and numerals to the quantities they represent?
- How accurately can students create a set of a given size? What kinds of counting errors do you notice? Do students double-check their count?
- How do students count a set of objects? Do they count each object once and only once? Do they skip some or count others more than once? Do they double-check their count? Do students organize the objects in any way to count them? Do they touch each object or move it from one place to another? Do they line them up? Do they have a way to keep track of which objects have already been counted?
- How do students represent quantities? Do they use pictures? tallies? numbers? words? objects? some combination of these?

Investigation 2: Taking Inventory

- Do students know the sequence of number names? Do they forget any of them? Which ones? Do they skip certain numbers or mix the order of the names? Which ones? Do they self-correct? Do they notice other students' mistakes?
- When counting a set of objects, do students count each object once and only once? Do they skip some or count others more than once? Do they double-check their count? Do they organize the objects in any way? Do they touch each object or move it from one place to another? Do they line up the objects? Do they have a way to keep track of which objects they have already counted?
- Do students represent quantities with pictures? numbers? words? objects? some combination of these? Which are they most comfortable with?
- How comfortable are students working with a partner? Do they work cooperatively? independently of each other? How do they handle and resolve disagreements?

Investigation 3: Comparing Towers

- What language do students use to describe and compare the lengths of two objects? Are they familiar with vocabulary such as *long* and *short? longer (higher, taller)* and *shorter? bigger* and *smaller?* How accurately do they use this vocabulary?
- How do students determine which is longer, an object or a tower of 10 cubes? Do they compare the object directly to the tower? Do they align the tower and object at one end? Do they stand materials next to each other on a flat surface to compare them?
- What language do students use to describe and compare quantities? How accurately do they use vocabulary such as *less* and *more? least* and *most? same* or *equal?*
- How do students compare amounts to determine which is greater? Do they use materials (for example, when looking at two numbers, do they build a tower of cubes for each number and then compare the height of the towers)? Do they count the quantities and compare the totals?

Investigation 4: Counting and Comparing

- How do students compare amounts to determine which is greater? Do they use materials? Do they count the quantities and compare the totals?
- What language do students use to describe and compare quantities? Do they use the terms *less* and *more? least* and *most? same* or *equal?*
- How are students counting and keeping track of quantities? Do they count each object once and only once? Do they double-check their count? Do they organize objects as they count? Do they have a way to keep track of which objects have been counted and which have not? How are their strategies developing and changing over time?

Investigation 5: Least to Most

- How do students order amounts (such as cube towers) from least to most? Do they begin with the tower that has the most or least cubes? Do they begin by grouping all same-size towers? Do they compare pairs of towers? Do they order the towers visually, building staircases with them? Do they order them numerically, using the number of cubes in each tower?
- How are the students counting and keeping track of quantities? Are their strategies developing and changing over time?

Investigation 6: Arrangements of Six

- How do students count out and arrange a set of six tiles? Does each tile in their arrangements touch another tile in some way? Can students explain how they arranged the tiles? Do they find ways that are easy to remember or explain visually? numerically? both?
- Do students realize there are many possible solutions to the problem? How comfortable are they with this? Do they find more than one arrangement on their own? at your suggestion?
- How are the students counting and keeping track of quantities? Are their strategies developing and changing over time?
- How are students representing quantities? Are their representations of their mathematical work and ideas changing over time?

PREVIEW FOR THE LINGUISTICALLY DIVERSE CLASSROOM

In the *Investigations* curriculum, mathematical vocabulary is introduced naturally during the activities. We don't ask students to learn definitions of new terms; rather, they come to understand such words as *triangle, add, compare, data,* and *graph* by hearing them used frequently in discussion as they investigate new concepts. This approach is compatible with current theories of second-language acquisition, which emphasize the use of new vocabulary in meaningful contexts while students are actively involved with objects, pictures, and physical movement.

Listed below are some key words used in this unit that will not be new to most English speakers at this age level, but may be unfamiliar to students with limited English proficiency. You will want to spend additional time working on these words with your students who are learning English. If your students are working with a second-language teacher, you might enlist your colleague's aid in familiarizing students with these words, before and during this unit. In the classroom, look for opportunities for students to hear and use these words. Activities you can use to present the words are given in the appendix, Vocabulary Support for Second-Language Learners (p. 112).

Note: While all students will be developing mathematical meaning for the following terms through the activities in this unit, English-speaking students will have heard and used the terms more generally in conversation, while limited English speakers will have had little or no experience with them. Second-language students will benefit from some prior exposure to these terms, to help them comprehend student-generated discussion as well as offer their own ideas related to comparison of quantities and lengths.

more, most, less, least, fewer, fewest, same Students use these terms as they count objects and compare the different amounts.

long, longer, longest; short, shorter, shortest Starting in Investigation 3, Comparing Towers, students begin talking about towers of cubes of different lengths and comparing lengths of other objects to a standard tower of 10 cubes.

higher, taller When comparing length on a vertical dimension, students often use the words *higher* and *taller* in place of *longer.*

Multicultural Extensions for All Students

Whenever possible, encourage students to share words, objects, customs, or any aspects of daily life from their own cultures and backgrounds that are relevant to the activities in this unit. For example:

- For Investigation 2, you might find items to put in the Inventory Bags that specifically reflect particular cultures in your classroom, such as sets of chopsticks.

- When you are comparing name lengths in Investigation 4, you might explore the naming traditions in different cultures. How many names are children given? Do they use nicknames? Are children named for particular relatives? Do names have a special meaning? Family members might be willing to share information about their cultures, or see *The Melting Pot Book of Baby Names* by Connie Lockhart Ellefson (Better Way Books, 1995), which includes some discussion of naming traditions around the world.

- Check your library for counting books with themes related to other cultures. *Count Your Way Through Mexico* by Jim Haskins (Carolrhoda Books, 1989) is one of a series of books that uses the numbers 1 through 10 and full-color illustrations to introduce a particular country and culture, including China, Japan, India, Korea, Israel, the Arab world, Russia, Brazil, Greece, Ireland, and many others. Also look for the following:

Emeka's Gift: An African Counting Story by Ifeoma Onyefulu (Cobblehill Books, 1995).

Moja Means One: Swahili Counting Book by Muriel Feelings (Pied Piper, 1976).

One White Sail: A Caribbean Counting Book by S. T. Garne (Green Tiger Press, 1992).

Ten Little Rabbits by Virginia Grossman (Chronicle Books, 1991) (Native American cultures).

Investigations

INVESTIGATION 1

Counting Books

Focus Time

Counting Books (p. 4)

To begin their exploration of counting, students read a counting book about the numbers 0 through 12. After examining the book to see how it is structured, they begin to make their own counting books, with one page for each number 0 through 6.

Choice Time

My Counting Book (p. 10)

Students finish making their individual counting books for the numbers 0 through 6.

Grab and Count (p. 12)

Students grab a handful of objects from a bin. Before counting, they estimate the amount, using a pile of five objects as reference. Then students find a way to record their handful using pictures, numbers, or words.

Counting Jar (p. 14)

Students count the objects in a clear jar. Then they use different materials to create collections with the same number of items.

Mathematical Emphasis

- Thinking about what, when, and why people count
- Recognizing numerals and number names
- Connecting numerals to the quantities they represent
- Developing strategies for counting and keeping track of quantities
- Creating a set of a given size
- Representing quantities with pictures, numerals, or words

Teacher Support

Teacher Notes

Counting Is More Than 1, 2, 3 (p. 16)

Students' Counting Books (p. 19)

Dialogue Boxes

What Do We Count and Why? (p. 17)

Pictures in a Counting Book (p. 18)

INVESTIGATION 1

What to Plan Ahead of Time

Focus Time Materials

Counting Books

- *Anno's Counting Book* by Mitsumasa Anno (HarperCollins Publishers, 1986) (optional; see p. 4 for alternatives)
- Eight sheets for The Counting Book, (pp. 115–122): 1 complete set for each student
- Stapler, or hole punch and string, yarn, or ribbon for binding books
- A wide variety of coloring and drawing materials, stencils, stickers, stamps and stamp pads, scrap materials, magazines and catalogs to cut from, scissors, tape, glue sticks

Choice Time Materials

My Counting Book

- Counting books and art supplies from Focus Time

Grab and Count

- Uniform items such as interlocking cubes, color tiles, teddy bear counters, or Styrofoam peanuts (sized so that a "handful" will be 5–12 objects), in open containers such as shoe boxes or plastic tubs
- Student Sheet 1, Grab and Count (p. 123): 2–3 per student
- Pencils, crayons, or markers

Counting Jar

- A clear container filled with countable objects such as small balls, blocks, or tiles
- Prepared recording sheet on chart paper (see p. 102)
- Paper plates: 1 per student

Family Connection

- Family letter (p. 114): 1 per family

Focus Time

Counting Books

What Happens

The class talks about what it means to count, and when and why people count. Together, the class reads *Anno's Counting Book*. Students make observations about each page and count the items there. They also talk about the way the book is organized, in preparation for making their own counting books. Their work focuses on:

- thinking about what, when, and why people count
- developing strategies for counting and keeping track of quantities
- recognizing numerals and number names
- connecting numerals to the quantities they represent
- representing quantities with pictures and numerals

Materials and Preparation

- Obtain a copy of *Anno's Counting Book* by Mitsumasa Anno, in big book form if possible, or a similar book such as *From One to One Hundred* by Teri Sloat (A Puffin Unicorn Book, 1991); *Ten Black Dots* by Donald Crews (Greenwillow Books, 1986); *Feast for 10* by Cathryn Falwell (Clarion Books, 1993); or *Count and See* by Tana Hoban (Macmillan, 1972).
- Have chart paper available (optional).
- Duplicate the eight sheets of The Counting Book (pp. 115–122) and use a stapler or a hole punch with string, yarn, or ribbon to bind one complete book for each student.
- Gather art supplies for the counting books, including colored pencils, markers, and crayons, stencils, stickers, stamps and stamp pads, magazines and catalogs to cut from, scissors, tape, and glue sticks.

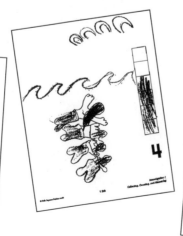

Activity

Anno's Counting Book

Note: This Focus Time is based on *Anno's Counting Book*. If you cannot obtain this title, it's easy to adapt the activities to an alternative book with pictures to count for each number, such as *From One to One Hundred*. (If you use this alternative, read only the pages through the number 10.)

We're going to read a counting book today. Who has an idea about what it means to count? Can you think about some times when we count in our class?

Some students will probably count for you, demonstrating *how* to count. Encourage them to think about *when* and *why* people count. Students might suggest activities such as taking attendance, serving snack, looking at the calendar, and so on. Ask them to think about what they are counting at these times and the reason for counting. See the **Dialogue Box**, What Do We Count and Why? (p. 17), for the ideas that emerged in one class.

The book we're going to look at today is called *Anno's Counting Book*. It was written and illustrated by a man named Mitsumasa Anno, who calls himself Anno when he writes and draws books like these. What do you think this book is going to be about? What do you notice about the cover? Does it give you any clues?

When students have had a chance to share their ideas, ask for a prediction about the first page.

If this is a counting book, does anyone have an idea about what might be on the first page? Why do you think so?

After hearing students' ideas, turn to the first page, the "0" page. Students may be surprised that the book does not start at 1, since people often begin counting with 1. (If the book you are using begins with 1, just start there.)

Is this what you expected? Most of us thought this page would be the page for the number 1. Is it? How do you know? What do you notice about this page?

Some students might recognize that this is not the page for the number 1, but not know what page it is. Others may recognize the numeral and know that it's the "0" page. Some might think it's the "1" page because there's one river. Others might notice the hills, the sky, the snow, the numeral, the column of squares, or the fact that there are no words or people—or anything else that appeared on the cover. If students do not bring it up, ask about the numeral on this page.

I noticed something over here. *[Point to the zero on the right-hand page.]* **Did anyone else notice this? Does anyone have an idea what it might be or what it means?**

Some students might think that the zero is the letter O. Some might say it's a circle. If no one is familiar with zero, explain that it is a number.

Zero is another number, just like 1, 2, and 3. How many of you have ever heard of zero? Does anyone know what a zero means?

Zero is a complicated concept that not all kindergarten students will be familiar with or ready to grasp. Some will understand that zero means nothing—for example, if you have zero apples, you don't have any at all.

The author of this book drew this picture to go with zero. But Alexa saw ten squares, and Maddy saw some hills, and Luke saw one river. Why do you think Anno drew this picture to go with zero?

Give students a chance to explain their ideas. Encourage them to keep thinking about this page, and explain that you will return to it after they have seen the rest of the book.

If this page is the zero page, what do you think will be on the next page? Why do you think so?

When students have made and explained their predictions, turn the page.

Is this what you expected? What do you notice about this page? Do you see just one of anything on this page?

Students are likely to disagree on some of the illustrations. For example, there are two trees on the page, but they are different kinds of trees. There are two people, but one is an adult and one is a child. Encourage students to use what they know about the book—its illustrations, cube representations, and the numerals on each page—to resolve these disputes. See the **Dialogue Box**, Pictures in a Counting Book (p. 18), for excerpts from the discussion of this book in one class.

Continue reading the book, encouraging students to comment on what they notice and to find groups of things that go with the number of the page. Occasionally ask students the name of the numeral or what number they think will be next.

As you continue through the book, students will probably begin to notice more and will want to go back and look for these new elements on earlier pages. Investigate these things as students notice them. There are many layers in this book and lots of detail. Teachers find that each time they read the book, students seem to find new things.

If students do not suggest going back to the zero page in light of other discoveries, return to it after you finish the book. Do students have any new ideas, now that they have seen the rest of the book, about what makes that first page the zero page?

Activity

Make-Your-Own Counting Books

If we wanted to make a counting book like Anno's, what sorts of things would we need to include on each page? What was special about the way Anno illustrated his book? What does every page in his book have?

Hold the book open and flip through the pages as students think about these questions. You might record students' ideas on chart paper, making simple sketches to illustrate them. Turn through the book to look for repeated examples of each suggested idea, being sure everyone understands it. To encourage students to listen to each other, ask them to think of characteristics of the book that someone else has not already shared. Your list might include ideas like these:

- On every page there are more people. They bring more things and build more houses.
- It's a counting book. Each page goes up one number: 1, 2, 3 . . .
- Every page has a big number on the side. The number is somewhere in the picture too, like on the clock or on a house.
- Every page has a cube tower. The number of cubes in the tower is the same as the number on the page.
- Every page shows several groups that represent the number. (On the "4" page there are four houses, four children, four adults, and so on.)

Explain that students are going to make their own counting books, from 0 through 6. Show them the blank counting books you have prepared for them to complete. Explain that they can draw and color their illustrations, but they can also use stamps, stickers, stencils, and items cut out of magazines or catalogs. Remind students of things they have noticed about Anno's book as you talk about what their own books might look like.

If there is time, students can get started on their counting books. Explain that they will come back to this activity during Choice Time, so they should not worry about having enough time to finish. Also let them know where you will store their books so they can find them when they are ready to work on them again.

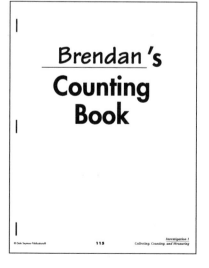

Start by helping students put their name in the blank on the cover. Observe as they begin work on their books. Keep in mind that creating a set of a given amount is a new task, different from counting a given set. You may discover that some students are more comfortable counting objects they can manipulate rather than groups that are static, such as pictures that cannot be moved around or picked up. See the **Teacher Note**, Counting Is More Than 1, 2, 3 (p. 16), for more information on counting in kindergarten.

As you circulate, try to keep the focus of this activity on the mathematics. If some students are spending too much time on detailed drawings, encourage them to draw one group, but to think about some quicker ways to create other groups, such as rubber stamps, stencils, stickers, or pictures cut out of magazines or catalogs. Such materials are also good for students who are not comfortable with drawing. The section About Assessment (p. I-8) alerts you to other special considerations when kindergartners are recording their ideas on paper.

Observing the Students

Consider the following as you watch students work on their Counting Books.

- How do students figure out how many items to draw on each page? Do they recognize the numeral on the page? Do they know what quantity it represents, or do they rely on knowing the counting sequence by rote to tell them what is next?

- How comfortable are students creating the groups of objects for each page? Do they show more than one group for each page? If they are illustrating the "2" page, do all the groups have two things in them? Do students color in an accurate number of squares?

- How accurate are students as they count? Do they say one number for each object? Do they count each object once, or do they count some more than once or skip some? Do they double-check their counts?

- Can students count out loud to 6 successfully? Do they forget certain number names? Which ones? Do they mix the order of the number names?

- How do students keep track while they are drawing? For example, while drawing five people, what do they do after they've drawn the third? Do they recount from 1 to see how many they have? Do they know they have three and go on to the fourth? Do they know they need two more?

See the **Teacher Note,** Students' Counting Books (p. 19), for a discussion of some of the difficulties you may have in interpreting students' work.

Focus Time Follow-Up

Seasons Use *Anno's Counting Book* to introduce and talk about the months and seasons of the year. Opening to any page of the book, ask students if they can tell what time of year it is. How do they decide? What clues do they use? Talk about how the living things in the book change as the seasons change.

Time You can also use *Anno's Counting Book* to talk about how we show time. Focus attention on the changes in the clock face from page to page.

Forming a Community As a connection to social studies, use *Anno's Counting Book* to explore the formation of a community. For example, if there were words in this book, what story would the book tell? When does this story take place? *Where* does it take place?

Family Connection Send home the signed family letter or the *Investigations* at Home booklet to introduce your work in this counting unit.

Three Choices If Choice Time is not already a standard part of your kindergarten program, refer to the **Teacher Note**, About Choice Time (p. 92), for more information.

For the first Choice Time of this unit, students continue their work on the counting books they started during Focus Time. You may also introduce two more activities that focus on counting: Grab and Count (p. 12), and Counting Jar (p. 14).

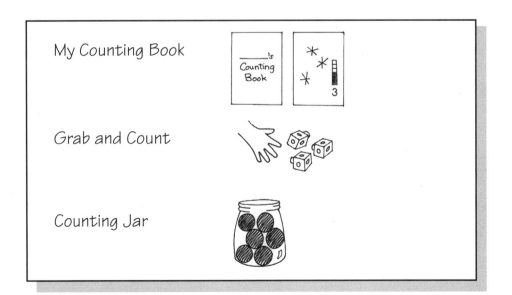

Choice Time

My Counting Book

What Happens

Students finish making their own counting books with pages from 0 to 6. Their work focuses on:

- developing strategies for counting and keeping track of quantities
- recognizing numerals and number names
- connecting numerals to the quantities they represent
- representing quantities with pictures and numerals

Materials and Preparation

- Make students' counting books available.
- Set out art supplies from Focus Time.

Activity

Students continue working on their own counting books, filling the pages in the style of *Anno's Counting Book*. Before students start on this Choice Time activity, you might want to recall as a group the special features of Anno's book. If you listed these on chart paper during Focus Time, you might leave this chart posted where students can refer to it.

As you observe students working, spend some time asking them to tell you about their drawings. The **Teacher Note**, Students' Counting Books (p. 19), illustrates some typical kindergarten work on this activity and the types of issues that may arise in your class.

Observing the Students

- How do students figure out how many items to draw on each page? Do they recognize the numeral on the page? Do they know what quantity it represents, or do they rely on knowing the counting sequence by rote to tell them what is next?

- How comfortable are students creating groups through six? Do they show more than one group of objects for each page? Do all the groups on a page have the same number in them? Do students color in the cube strip with an accurate number of cubes?

- How accurate are students as they count? Do they say one number for each object? Do they count each object once, or do they count some more than once or skip some? Do they double-check their counts?

- Can students count out loud to 6 successfully? Do they forget certain number names? Which ones? Do they mix the order of the number names?

- How do students keep track while they are drawing? If they are in the process of drawing five people, what do they do after they've drawn the third? Do they count from 1 to see how many they have? Do they know they have three and move on to the fourth? Do they know they need two more?

Variations

- If some students finish early, encourage them to add more examples of groups to each of their pages through the number 6.

- Some students may want to extend their counting books to show larger numbers. You can create additional pages for the numbers 7 through 12 for these students. Add a second tower of six cubes to each page, rather than extending a single tower up to 12.

Choice Time

Grab and Count

What Happens

Students grab a handful of objects. They count to find out how many objects they have. Then they make a representation that tells about their handful. Their work focuses on:

- developing strategies for counting and keeping track of a set of objects
- representing quantities with pictures and numerals

Materials and Preparation

- Set out open containers of uniform items such as interlocking cubes, color tiles, teddy bear counters, or Styrofoam peanuts. Choose objects that will result in handfuls of a reasonable size for your students to count (probably in the range of 5–12). Small items such as beans, buttons, or paper clips will result in quantities that are too large.
- Duplicate Student Sheet 1, Grab and Count (p. 123), 2–3 per student.
- Provide pencils, crayons, or markers.

Activity

Note: Students need to have freely explored any material you are using for this activity. If they have not done so previously, plan some time for this. See the **Teacher Note,** Materials as Tools for Learning (p. 95).

Introduce this activity by asking what it means to grab a handful of something. Once students have shared their ideas, show them a container with a material, such as the cubes, for which "a handful" will be a fairly small quantity. Count out five of that item and set these out for students to use as a reference in predicting quantities.

How many cubes do you think Renata can grab in one handful?

Take some predictions, and then ask your volunteer to grab a handful of cubes and place them where everyone can see them.

About how many do think there are? Does it look like five? more than five? less than five?

If the handful had more than five, count out five and ask the same questions about the amount left.

Is this more than five? less than five?

One goal is for students to make estimates and adjust them based on new information. Another goal is for students to get a really good sense of five—what it looks like and how to count it. Counting by ones, count all the items in the handful as a class, or have several volunteers count them.

Show Student Sheet 1, Grab and Count.

Renata is going to show her handful on this paper. What might she do so that someone looking at her paper could tell what she grabbed?

Students might suggest drawing a picture or writing the number or name of the items. They will likely generate other ideas for recording as they actually do the activity. Rather than demonstrating possible methods of recording, let students to generate their own.

Observing the Students

Consider the following as you watch students work on Grab and Count.

- As students count, do they know the sequence of number names? Do they forget or mix the order of some of the numbers? Which ones?

- How do students count their handfuls? Do they count each item only once? Do they skip some or count some more than once? Do they double-check their count?

- Do students count the items in a random configuration, or do they organize them in some way? Do they touch or move each item as they count it?

- How do students represent their handfuls? Do they draw each item? Do they use numerals? Do they arrange the objects in groups and represent them that way? Do they use tallies?

    Encourage students who are having trouble to try drawing pictures to show their thinking. However, keep in mind that for some students, this will prove difficult and frustrating. Don't hesitate to take dictation from students who are struggling to represent their work.

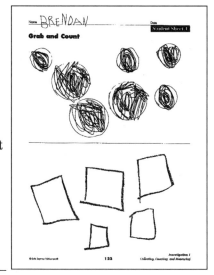

Note: The activity Grab and Count appears in four different forms in this unit. See the **Teacher Note**, Grab and Count and Its Variations (p. 91), for an overview. You can also vary the basic activity as suggested below.

Variation

- Introduce different materials, such as marker caps, rocks, and keys. Some students might be ready to try smaller materials, such as buttons, bread tabs, pebbles, or bottle tops, that will result in larger quantities.

Choice Time

Counting Jar

What Happens

In an activity that you may be doing as an ongoing classroom routine, students count the items in a clear jar to determine how many there are. They then create another collection that has the same number of objects in it. Their work focuses on:

- developing strategies for counting and keeping track of a set of objects
- creating a set of a given size

Materials and Preparation

- Obtain a clear plastic jar at least 6 inches tall and 4–5 inches in diameter. Place in the jar a number of items to be counted, such as small balls, cubes, or tiles. Choose a number that is manageable for most students in your class, probably in the range of 5–12.
- Gather a variety of countable materials, such as cubes, tiles, teddy bear counters, buttons, keys, lids, rocks, or bread tabs. There should be more of each material than there are items in the counting jar.
- Prepare a recording sheet on chart paper; see example on p. 102.
- Provide paper plates, 1 per student.

Activity

If you are using the full-year *Investigations* curriculum, you may have established Counting Jar as a regular classroom routine, as described on p. 102. If this is the case, the activity will need no introduction. Otherwise, introduce it by gathering students around the partially filled jar.

This is a special kind of jar—a counting jar. What do you think we will do with it? How could we find out how many [balls] are inside?

Ask several volunteers to count the objects in the jar. Encourage students to watch each other as they count and see what they notice about each person's strategy for counting.

Students might make a variety of mistakes, such as not always saying one number for each object counted. Because of this, students may not agree on how many items are in the jar. If discrepancies arise, focus the discussion on strategies for counting and for keeping track of the count, rather than on which person counted correctly. Involve students in thinking about reasons for the differences and reasons for losing count.

You might also use this opportunity to talk about double-checking as a tool for getting an accurate count. The **Dialogue Box**, Is It 10 or 11? (p. 33), shows how one teacher handled a counting discrepancy.

I'm going to put a new set of objects in the counting jar. When you visit the jar, your job will be to find out how many things are in it.

Show students the recording chart you have prepared. Explain how you would like them to record their counts—perhaps with their name and the number of items they think are in the jar. Point out the number line on the bottom of the recording chart.

Suppose I counted and I thought there were six things in the jar, but I didn't know how to write a 6. How could this chart help me?

If no one points to the numeral 6, show them how to count on the number line, beginning with 1, to find a particular number. For more on how young students learn to write numerals, see the **Teacher Note**, Observing Students As They Count (p. 32).

After you have counted what's in the jar, pick another kind of object. It could be any of the things you see here. Use those objects to make a group that has the same number of things as the jar. So if the jar has six things in it, you will make another group of six with these objects. Put your set of six on a paper plate *[demonstrate]*, **and label the new set with your name.**

Observing the Students

As students count the items in the jar and create a like-numbered set, circulate to get a sense of how comfortable they are with the task.

- Do students know the sequence of number names? Do they forget or mix the order of some of the numbers? Which ones?

- How do students count the objects in the jar? Do they count each item only once? Do they skip some or count some more than once? Do they arrange or organize the objects in any way? Do they touch or move each item as they count it? Do they double-check their count?

- How do students go about creating a group with a particular number of objects? Do they create sets of objects that have the same amount as the jar? Do they double-check their counts? Do they compare their new set to the set of objects in the jar?

Teacher Note: Counting Is More Than 1, 2, 3

Counting is the basis for understanding our number system and for almost all the number work in the primary grades. It involves more than just knowing the number names, their sequence, and how to write each numeral. While it may seem simple, counting is actually quite complex and involves the interplay between a number of skills and concepts. For more on what to expect as kindergartners learn about counting, see the **Teacher Note**, Observing Students As They Count (p. 32).

Rote Counting Students need to know the number names and their order by rote; they learn this sequence by hearing others count and by counting themselves. However, just as saying the alphabet does not indicate that a student can use written language, being able to say "one, two, three, four, five, six, seven, eight, nine, ten" does not necessarily indicate that students know what those counting words mean. Students also need to use numbers in meaningful ways if they are to build an understanding of quantity and number relationships.

One-to-One Correspondence To count accurately, a student must know that one number name stands for one object that is counted. Often, when young children first begin to count, they do not connect the numbers in the "counting song" to the objects they are counting. Children learn about one-to-one correspondence through repeated opportunities to count sets of objects and to watch others as they count. One-to-one correspondence develops over time, with students first counting small groups of objects (up to five or six) accurately, and eventually larger groups.

Keeping Track Another important part of counting accurately is being able to keep track of what has been counted and what still remains to be counted. As students first begin to count sets of objects, they often count some objects more than once and skip other objects altogether. Students develop strategies for organizing and keeping track of a count as they realize the need and as they see others use such strategies.

Connecting Numbers to Quantities We use numbers both to count a set of objects and to describe the quantity of those objects. Many young students are still coordinating these two aspects of number—the *ordinal* sequence of the numbers with the *cardinal* meaning of those numbers. In other words, we get to 5 by counting in order, 1, 2, 3, 4, 5. In this sequence, 4 comes after 3, and 5 comes after 4. Understanding this aspect of number is connected to the one-to-one correspondence between the numbers we say and the objects we are counting. However, being able to count accurately using this ordinal sequence is not the same as knowing that when we are finished counting, the final number in our sequence tells the quantity of the things we have counted.

Conservation Conservation of number involves understanding that three is always three, whether it's three objects pushed or linked together, three objects spread apart in a line, or some other formation. As students learn to count, you will see many who do not yet understand this idea. They think that the larger the arrangement of objects, the more objects there are. Being able to conserve quantity is not a skill that can be taught; it is a cognitive process that develops as children grow and develop. This unit provides many opportunities for kindergartners to bump up against this important developmental milestone.

Counting by Groups Counting a set of objects by equal groups, such as 2's, requires that each of the steps mentioned above happens again, at a different level. First, students need to know the 2's sequence (2, 4, 6, 8 . . .) by rote. They need to realize that one number in this count represents two objects, and that each time they say a number they are adding another group of two to their count. Keeping track while counting by groups becomes a more complex task as well. Students begin to explore counting by groups in the data unit *Counting Ourselves and Others* as they count the number of eyes in their class (counting by 2's). However, most students will not count by groups in a meaningful way until first or second grade.

DIALOGUE BOX

What Do We Count and Why?

The teacher begins this investigation of counting by asking the class to suggest activities and situations in which people count. The goal of the discussion is to encourage students to think of counting as more than just naming a string of numbers.

Who has an idea about what it means to count?

Alexa: Counting is the numbers.

Ida: Yeah, like 1, 2, 3, 4, 5.

OK, those are the numbers you say when you count. But why do people count? What are they doing when they count?

Henry: At morning meeting we count.

Ravi: You count all the kids. You pick a name and that person starts. You say your number and go around the circle. Let's say you pick my name out. I say "1" and Justine says "2," and the next person says "3," and then "4" . . .

Henry and Ravi said we count during attendance. Can anyone think of some other times that we count during the day?

[The class brainstorms a list that includes attendance, snack, calendar, and the chart of number of days in school.]

Ravi said earlier that we count people at attendance time. Why do you think we do that? Why do we need to count the kids in our class?

Maddy: Because you have to tell the office. How many kids are here and how many kids didn't come to school. So they know.

What about snack time? Gabriela said we count at snack time. What do we count at snack? Do we count kids?

Kylie: We count how many kids want chocolate milk and how many want white milk.

Renata: When I was snack person, I had to count and give everyone at my table five crackers so everyone would have the same.

So were you counting people?

Renata: No, I was counting snack. The crackers.

Why was it important to count how many crackers?

Renata: Because what if someone got too many? That wouldn't be fair. Or if I messed up and didn't give someone five, they might feel bad that they got not so much as everyone else.

Teacher Support ■ 17

DIALOGUE BOX

Pictures in a Counting Book

This class is reading *Anno's Counting Book*. Confusion arises as students notice that on the page for the number 1, there are two people and two trees. Students grapple with issues of classification as they investigate how the illustrations represent the numeral on the page.

Tarik thinks this is the "1" page. Why do you think Tarik has that idea?

Thomas: Because one family moved in and there was no one before.

Maddy: Because there's a 1 here *[points to the numeral]*, and a 1 here on the snowman.

Felipe: And one cube, too.

Justine: It's a counting book. After zero comes one.

If this is the "1" page, do you see one of anything on this page?

Justine: Sure, there's one tree.

[Other students offer one house, one river, one bird, one sun, one snowman, one shovel, one dog.]

Felipe: I think it's the "1" page, but . . . I see two trees. And two people, too. See? There's one skiing and there's one by the snowman.

Interesting! We think this is the "1" page, but Felipe sees two trees *[points them out]* **and two people** *[points]*. **What do others think about that?**

Thomas: Well, everything's not one. There's two chimneys and lots of windows . . .

Justine: I didn't see the other tree. But they look different. They're not the same trees.

Tarik: I think so, too. It's the same, like there's two animals, but there's *one* dog and *one* bird. There's two trees, but they're different.

What about the people? How are they "1"?

Tarik: It's kind of the same thing. There's two people, but one looks like a kid and the other's a grown-up.

Oscar: Or maybe it's a mom or dad and it's one, one family.

As these students think about how the illustrations on each page represent the numeral on the page, they are beginning to connect numerals with the quantities they represent.

18 ■ *Investigation 1: Counting Books*

Students' Counting Books

Teacher Note

Observing students as they work and listening to them explain what they are doing is especially important in kindergarten, when students' written work can be difficult to decipher and understand. In one classroom where students were making their own counting books based on *Anno's Counting Book*, the teacher found that her observations of and conversations with students were essential to interpreting their written work.

The teacher provided a variety of materials for students to use in making their books. Rubber stamps and shapes punched from colored paper (to be glued down) offered students an alternative to drawing, which was a difficult task for many.

Finding an Underlying Order At first glance, the teacher was concerned about Maddy's work (below). Her "2" page and "3" page looked chaotic; it was hard to imagine that Maddy had counted correctly. However, upon closer inspection and an explanation from Maddy, the teacher saw that, in fact, Maddy had included many groups of the correct number on each page, often identified by color (two red bears, two green bears, two green rabbits, two blue dinosaurs, and so forth). Her groups were just difficult to recognize because of the scattered placement and the "extra decorations."

Continued on next page

 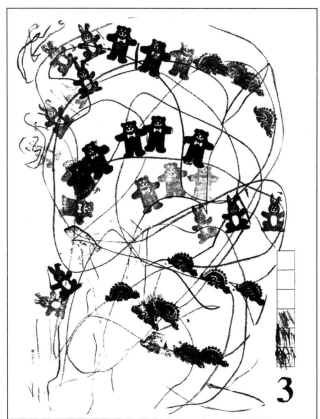

Two pages from Maddy's Counting Book

Teacher Note *continued*

Clarifying a Student's Intent Carlo used a variety of materials, including rubber stamps (bears, bunnies, and dinosaurs), drawings (surfers and pumpkins), and words (*boo* and *boom*) to illustrate his book. He began with a page similar to Anno's first page: his "0" page shows only a river and a bridge. However, this seems to have confused Carlo because when he got to the "1" page, he had already drawn one bridge and now went on to draw two bridges. To add to the confusion on his "1" page, Carlo seemed to be using rubber stamps in groups of three. But this was clarified when the teacher asked him to explain his thinking and Carlo said, "There's one red bear, one red bunny, one red dinosaur. There's one blue bear, one blue bunny . . . [and so forth]."

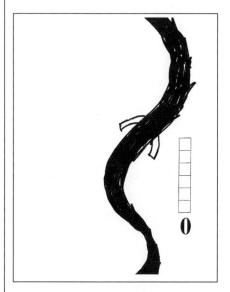
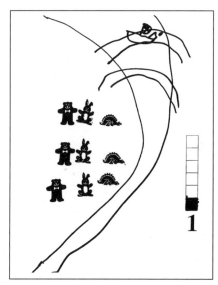
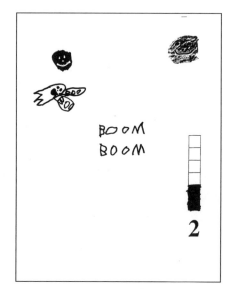
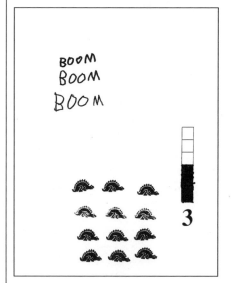

Six pages from Carlo's Counting Book

Teacher Note *continued*

Correcting an Error Tess also used a variety of materials to illustrate her book, including colored markers and shapes punched from paper. Her pictures were accurate and clearly organized. However, after illustrating the "6" page, Tess was concerned. She brought her book to the teacher and explained the dilemma—she had accidentally made a group of more than six objects. When asked how she might fix her mistake, Tess had just one idea: crossing out two of the objects, a solution that did not please her. The teacher suggested that Tess might add more things to each group (more cutout shapes to the top group and more hand-drawn hearts below), resulting in two groups of six. Tess was unable to make sense of this idea and decided to cross out two of her hearts, leaving a single group of six mixed objects.

Interpreting Drawings For Jacob (below), drawing is an arduous task. Because he does not use representational drawings, it is difficult to interpret his work, and at first glance his work may seem inaccurate. But when Jacob sat down to talk about his book with the teacher, it was clear that the pictures had meaning for him, and he was able to point out the groups that represented the number on each page. In this instance, as in the previous three, a brief conversation helped the teacher assess the students' understanding.

Two pages from Tess's Counting Book

 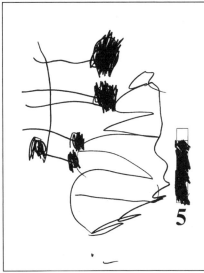 

Three pages from Jacob's Counting Book

Teacher Support ■ **21**

INVESTIGATION 2

Taking Inventory

Focus Time

Taking Inventory (p. 23)

In a whole-group meeting, the class discusses "taking inventory" as a reason for counting things. Students then work in pairs to inventory the contents of a bag holding several similar items. After they count the items, they make a representation to show their results. Students then meet again as a whole group to share the different ways they found to record and to discuss strategies for counting and for keeping track as they count.

Choice Time

Inventory Bags (p. 30)

Students continue the Focus Time activity, counting the items in inventory bags and making representations of their work.

Continuing from Investigation 1
My Counting Book (p. 10)
Grab and Count (p. 12)
Counting Jar (p. 14)

Mathematical Emphasis

- Connecting numerals to the quantities they represent
- Developing strategies for counting and keeping track of quantities
- Creating a set of a given size
- Representing quantities with pictures, numerals, or words

Teacher Support

Teacher Notes

Observing Students As They Count (p. 32)

From the Classroom: Students' Difficulties with Counting (p. 34)

Dialogue Box

Is It 10 or 11? (p. 33)

INVESTIGATION 2

What to Plan Ahead of Time

Focus Time Materials

Taking Inventory
- For the inventory bags, classroom materials to make sets of 5–12 identical or similar items, such as erasers, pencils, buttons, paper clips, rulers, crayons, markers, cubes, tiles, teddy bear counters, pattern blocks, thread spools
- Small paper bags: 1 per pair plus extras. Label each bag with a letter and the name of the contents.
- Student Sheet 2, My Inventory Bag (p. 124): 3–4 per student (includes supply for Choice Time)
- Coloring and drawing materials, stencils, stickers (including dot stickers), rubber stamps and stamp pads, tape, glue sticks

Choice Time Materials

Inventory Bags
- Inventory bags and supply of Student Sheet 2, from Focus Time

My Counting Book
- Students' counting books and art supplies from Investigation 1

Grab and Count
- Uniform items such as interlocking cubes, square color tiles, teddy bear counters, or Styrofoam peanuts, in open containers
- Student Sheet 1, Grab and Count (p. 123): copies remaining from Investigation 1

Counting Jar
- The Counting Jar and your prepared recording sheet
- Paper plates: 1 per student
- Collections of countable materials

Investigation 2: Taking Inventory

Focus Time

Taking Inventory

What Happens

The class counts the contents of one inventory bag as a group. Then student pairs each take one bag of items to inventory on their own. After counting the items in their bag, students each find a way to represent the contents of that bag. Students meet to share their strategies, both for counting and for recording their results. Their work focuses on:

- developing strategies for counting and keeping track of a set of objects
- representing quantities with pictures, numerals, or words
- working with a partner

Materials and Preparation

- Make inventory bags, 1 per pair plus extras, by filling small bags with 5–12 identical or similar classroom items. The number in the bags will depend on how high your students are counting comfortably. Place a different kind of item in each bag. Label each bag with a letter and the name of the items inside. Possible items include erasers, pencils, buttons, paper clips, rulers, crayons, markers, cubes, tiles, teddy bear counters, pattern blocks, and thread spools.
- Duplicate Student Sheet 2, My Inventory Bag (p. 124), 3–4 per student (includes supply for Choice Time).
- Gather a variety of materials for making representations, such as coloring and drawing materials, stencils, stickers (including dot stickers), rubber stamps and stamp pads, and tape or glue sticks.

Activity

Counting What's in the Bag

Show students one of the inventory bags.

Today I thought of another time when I count. We have lots of different materials in this classroom and some of them are things that we use up, like pencils. I count those kinds of things so that I know when we are running out and I need to order more.

Lots of people do this kind of counting. It even has a special name. When we count to find out how many of something we have in a classroom or a store, or in our house, it's called "taking an inventory." That means we count to find out how many of something there is.

Today we're going to take an inventory of some of the materials in our classroom. I made bags that hold the things you'll be counting.

Show students the contents of one inventory bag. With a student volunteer as your partner, count the items. Since it is likely that pairs of students will get different numbers when they count, you might ask the volunteer to count first and then come up with a different number yourself. That way, you can model a way to disagree respectfully, and you can talk about what to do when this happens.

Discuss strategies for resolving such discrepancies, such as having each person carefully count the objects again while the other watches and listens closely, or counting the objects in a different way to double-check. Encourage students to work together to count the objects cooperatively, but acknowledge that sometimes it will be too hard for partners to come to an agreement. When this happens, students' representations should show how many they, personally, think are in the bag. There will be time to discuss discrepancies after the inventories are completed.

Each pair will get a bag, and together you'll need to figure out how many objects are inside. Then you'll each make your own representation, or picture, that shows what kind of things are in the bag and how many things there are.

Talk briefly about how students might use Student Sheet 2, Inventory Bag, to represent the contents of the sample bag. (Students who have done the Choice Time activity Grab and Count have already had some experience with representing a set of objects on paper.)

Focus Time: Taking Inventory ■ **25**

If you were to make a representation for this bag, what might you put on this sheet to tell someone what you found out? What's the important information about this bag?

When students have shared their ideas, summarize what a representation of these data might show: the letter on the bag, what kind of object was in the bag, how many objects there were. Allow students to generate their own ways of recording, rather than demonstrating possible methods.

Student partners take a bag and two copies of Student Sheet 2. They find a place to work and count the items in their bag. Then they use the student sheet to show the class what they found out.

As students work, circulate to observe how they are counting and representing the materials in the bags. For more information on students' issues with counting, see the **Teacher Notes**, Observing Students As They Count (p. 32) and From the Classroom: Students' Difficulties with Counting (p. 34).

Observing the Students

Consider the following as you watch student pairs taking inventory.

- Do students know the sequence of number names? Do they forget any of the names? Which ones? Do they skip certain numbers or mix the order of the names? Which ones? Do they self-correct? Do they hear mistakes in the counting sequence if someone else makes them?

- Do students count each object only once, or do they count some more than once and skip some altogether? Do they double-check their counts?

- Do students organize the objects in any way to count them? Do they touch or move each object? Do they line up the objects? Do they have a way to keep track of which objects have already been counted?

- Do students notice if they get two different counts for the same group of objects? How do they respond when this happens? Do they double-check the count? count the group a different way? ask a partner to double-check?

- How do students represent their inventory on paper? Can you tell from their representation which bag they had? what material was in the bag? how many objects were in the bag?

- How do pairs of students share this task? Do they work cooperatively or independently? How do they handle disagreements?

This student used only pictures to represent his inventory bag, tracing the buttons he counted. What he counted and how many there were (12) is very clear.

This student counted 5 bread tags in her inventory bag. To represent her work, she drew a picture of each tag and labeled her paper with the total, 5.

In the first part of the kindergarten year, many students will not yet say one number name for each object they count and will make a variety of counting mistakes. Because of this, many pairs will disagree on how many items are in their bag. As you circulate and observe students, take note of pairs that disagree. Encourage them to use the strategies you brainstormed as a class for handling disagreements, particularly if they are counting the objects only once. At this age, many discrepancies will not be resolved. You can use these cases as examples during the follow-up discussion.

Some pairs may finish early. These pairs can trade bags, do another inventory, and then compare their results with one another. If no one is available to trade bags, pairs who finish early can inventory one of the extra bags you prepared.

Activity

What Did You Find Out?

When students have had a chance to inventory and represent at least one bag, call them together to share their work. Ask them to bring their student sheets with them. For more information on holding this type of discussion, see the **Teacher Note**, Encouraging Students to Think, Reason, and Share Ideas (p. 96).

Start by focusing the discussion on different methods of representing the inventory information. Ask a volunteer to show and explain his or her work to the class.

Tarik showed the information about his bag by drawing a picture of each object in the bag. Did anyone else draw a picture of each item in the bag? Raise your hand if you did. Did anyone show the information in a different way? I see that Charlotte used a number. That's a different way. Did anyone else use numbers? Raise your hands.

By asking for a show of hands for each method discussed, you can acknowledge everyone's work without taking the time for each child to share individually.

28 ■ *Investigation 2: Taking Inventory*

In this discussion or at a later time, ask students about any discrepancies that arose in their counts.

What happened if you counted and got one number and then your partner counted and got a different number? Does that matter?

Focus this discussion on strategies for counting and keeping track of or organizing the count. Students will have a variety of ideas about why different people came up with different numbers. Encourage them to think about possible reasons and to avoid "Because I know I'm right" kinds of answers. See the **Dialogue Box**, Is It 10 or 11? (p. 33), for the discussion that took place in one class over a counting discrepancy.

Focus Time Follow-Up

Four Choices In addition to continuing their work with inventory bags, students continue working on the three Choice Time activities from Investigation 1: My Counting Book (p. 10), Grab and Count (p. 12), and Counting Jar (p. 14). If your students have completed the work on counting books or Grab and Count, you might consider one of the variations listed for those activities.

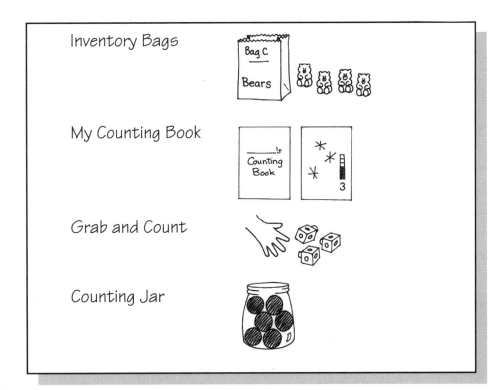

Choice Time

Inventory Bags

What Happens

Students continue to count the materials in inventory bags. For each bag they inventory, they make a representation that shows what was in the bag and how many there were. Their work focuses on:

- developing strategies for counting and keeping track of a set of objects
- representing quantities with pictures, numerals, or words
- working with a partner

Materials and Preparation

- Make available the inventory bags from Focus Time and the remaining supply of Student Sheet 2, My Inventory Bag. Also provide a variety of drawing materials and art supplies for making representations.

Activity

Introduce this choice by reviewing what students did with the inventory bags during Focus Time. You might display one student's work on Student Sheet 2 as a reminder of the process.

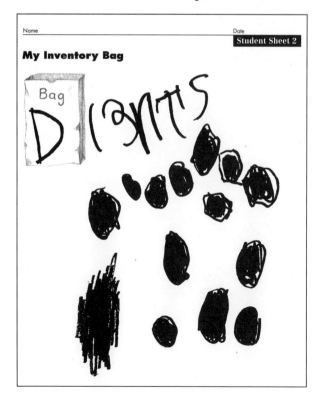

This student used pictures, numbers, and words to show that he counted 13 nuts. He drew a circle for each nut and wrote both the number and the word—"13 nts." When he recounted his circles to check, he realized he had one too many, so he scribbled out one circle.

We are going to continue to inventory the materials in our classroom. Everyone has already done at least one bag. Now you can choose any bag you like. When you have a count for the things in your bag, record your findings on the inventory bag sheet. For each bag you inventory, make your own recording to show how many are in the bag and what the material is.

Observing the Students

Consider the following as you watch student pairs work on Inventory Bags.

- Do students know the sequence of number names? Do they forget any of the names? Which ones? Do they skip certain numbers or mix the order of the names? Which ones? Do they self-correct? Do they hear mistakes in the counting sequence if someone else makes them?

- Do students count each object only once, or do they count some more than once and skip some altogether? Do they double-check their count?

- Do students organize the objects in any way to count them? Do they touch or move each object? Do they line up the objects? Do they have a way to keep track of which objects have already been counted?

- Do students notice if they get two different counts for the same group of objects? How do they respond when this happens? Do they double-check the count? count the group a different way? ask a partner to double-check?

- How do students represent their inventory on paper? What information do they show? Can you tell which bag they had? what material was in the bag? how many objects were in the bag?

- How do pairs of students share this task? Do they work cooperatively or independently? How do they handle disagreements?

Variation

Ask all the students in your class to inventory one or two bags that you specify. Post all the representations for a particular bag in one spot. Discuss the variety of ways students chose to represent the contents of the same bag. If students have found different totals for the same bag, that could spark an interesting debate.

Teacher Note: Observing Students As They Count

In kindergarten, you can expect to see a wide range of number skills within your class. Students in the same class can vary considerably in age and in their previous experience with number and counting. (For more on the concepts your students will be exploring as they count, see the **Teacher Note**, Counting Is More Than 1, 2, 3, p. 16.)

Your students will have many opportunities to count and use numbers in this unit and throughout the year. You can learn a lot about what your students understand by observing them. Listen to students as they talk with each other. Observe them as they count orally, as they count objects, and as they use numerals to record. Ask them about their thinking. You may observe some of the following:

Counting Orally By the end of the year, most kindergarten students will have learned to count by rote up to 10 and beyond, with some able to count as high as 100. Many will be able to count orally much higher than they can count objects. For many students who have learned the internal counting pattern or sequence (1, 2, 3 . . . 21, 22, 23 . . .), the "bridge" numbers into the next decade (such as 19, 20, or 29, 30) remain difficult. You may hear children count "twenty-eight, twenty-nine, twenty-ten." Just as the young child who says "I runned away" understands something about the regularities of the English language, the student who says "twenty-ten" understands something about the regularity of counting quantities. Students will gradually learn the bridge numbers through repeated experiences with counting and listening to the counting sequence.

Counting Quantities Most kindergartners will end the year with a grasp of *quantities* up to 10 or so. Some students may accurately count quantities above 10 or even 20; others may not consistently count quantities up to 5 or 6. Some students may be inconsistent—successful one time, having difficulty the next. Even when students can accurately count the objects in a set, they may not know that the last number they said also describes the number of objects in the set.

You may observe students who successfully count a set of cubes, but have to go back and recount the set to answer the question, "How many cubes are there?" These students have not yet made the connection of the counting numbers with the quantity of objects in a set. Students develop their understanding of quantity through repeated experiences organizing and counting sets of objects. In kindergarten, many of the activities that focus on quantity can be adjusted so that students are working at a level of challenge appropriate for them.

Organizing a Count Some students may be able to count objects they can pick up, move around, and organize with far more accuracy than they can count static objects, such as pictures of things on a page. You may observe some students who can count objects correctly when the group is organized for them, but you'll see others who have trouble organizing or keeping track of objects themselves. They will need many and varied experiences with counting to develop techniques for counting accurately and for keeping track of what they are counting.

Writing Numbers Knowing how to write numerals is not directly related to counting and understanding quantity; however, it is useful for representing a quantity that has been counted. Young students who are learning how to write numerals frequently reverse numbers or digits. Often this is not a mathematical problem but a matter of experience. Students need many opportunities to see how numerals are formed and to practice writing them. They should gain this experience by using numbers to record mathematical information, such as the number of students in school today or the number of objects on a page of a counting book. Numeral formation is related to letter formation; both are important in order to communicate in writing. We recommend that rote practice of numeral writing be part of handwriting instruction rather than mathematics.

DIALOGUE BOX
Is It 10 or 11?

During work with Inventory Bags, situations come up repeatedly in which different students count the same collection and arrive at different numbers. This is tricky for a teacher to handle, because clearly only one count is right. However, simply verifying which one is right may not improve students' understanding of numbers and counting. In fact, without probing further, you can't know whether the student with the more accurate count does in fact have a better understanding.

When discrepancies arise, focus on strategies for keeping track of a count, reasons for losing count, and ways to double-check. Involve students in thinking about why there might be differences in the count for a particular bag.

This class is sharing what students found while doing the Inventory Bags activity for the first time.

Who did an inventory of Bag F?

Tiana: Me and Xing-Qi did. We got pencils in our bag and there were 11.

Charlotte: But me and Ida did pencils, and there were 10, not 11.

Interesting. Two different pairs inventoried this bag. Tiana and Xing-Qi counted 11 pencils, and Charlotte and Ida counted 10. What do other folks think about that?

Shanique: Maybe some fell out of the bag or got lost.

That's certainly one reason they could have ended up with different numbers. What if I told you I'm sure that all the pencils were there for both pairs? None were lost and none were added to the bag. Are there any other reasons they might have gotten different numbers? . . .

I know some people had a hard time agreeing on how many things were in their bag. Luke and Ravi, I saw you having a similar problem. Can you tell us about what happened with your cube inventory?

Luke: Well, I counted first and got a bunch of numbers. Then I counted and got 9 cubes. And then Ravi counted them and he got 10.

Ravi, how did you and Luke work this out?

Ravi: We both counted the cubes again. And then I got 9 that time.

So no one took any cubes away, or added any cubes to their pile, but Luke and Ravi still got different numbers. Why do you think that might be?

Luke: Well, one time when I watched, Ravi touched a cube two times by mistake. And when I was counting at first I kept getting messed up 'cause I'd forget which ones I counted and I kept having to start over. Then I snapped them together in a line as I said the number and counted them that way. Then I got 9 two times in a row and so did Ravi.

So it sounds like snapping them together as you counted helped you keep track of the cubes, and that double-checking helped you make sure that there really were 9, not 10 or 8.

The teacher then returned the discussion to the bag of pencils, asking students for ways the two pairs could check their results.

You can reassure students that, in fact, it's quite easy for anyone to be off by one or two in a count. It's a familiar experience for most of us to recount something and get a different number. Students should understand that although it is easy to miscount, there are strategies that help us be sure of our result. These include recounting ourselves, asking someone else to count to double-check, and carefully touching each object or moving it from one spot to another as we say each successive number.

FROM THE CLASSROOM

Teacher Note ▸ *Students' Difficulties with Counting*

After introducing the Inventory Bag activity, I circulated among pairs of students to see what strategies they were using to count. I was amazed and somewhat overwhelmed at the variety of different ways students had for counting a group of objects.

Counting Pencils When I joined Binh and Wyatt, they were just pouring the pencils from their inventory bag onto the table. They seemed excited to find out what was in the bag. When I asked Binh to count how many pencils there were, he began pointing in the air and saying the number sequence. He seemed to know the sequence quite well (he counted up to 15), but did not seem to attach these words to the objects in the bag.

Wyatt then took a turn counting the pencils. He counted accurately, lining up pencils and touching each one as he counted it. He paused when he hit 10, looking to me for the name of the next number. His partner offered, "11." He announced that he thought there were 12 pencils. Either he didn't remember that Binh had said 15, or that fact didn't bother him.

Counting Crayons When I joined Channary, she was already hard at work on her representation. The crayons from her inventory bag were on the table, sorted by color. She had made detailed drawings of two blue crayons. I asked how many crayons were in her bag, and she surprised me by answering, "I don't know yet." I realized that she was drawing the crayons one by one. She would put a crayon on her paper, draw it, put it back in the bag, and then take the next crayon from the table to be counted. Her focus was on the drawing, not the number of crayons.

When I revisited Channary's table, she had completed a picture of each crayon in her bag. The crayons were drawn on the paper haphazardly, where they would fit. Channary had a hard time counting them because they were drawn almost in a circle. Her partner, Javier, looked up from his own work to suggest that she count one crayon at a time and place them in a line on the table. With his help, Channary was able to count the crayons.

Counting Cubes When I joined Devon and Jamilla, they had already written "8 cubes" on their papers, using the label on their bag to spell *cubes*. I asked them to show me how they found out there were eight.

Devon counted the cubes, which were in a pile on the table. She touched the cubes as she counted them, but did not move them in any way—they remained in a scattered pile on the table. As she counted, I noticed that she skipped several cubes and recounted several others, but managed to end up with eight.

Jamilla was carefully observing Devon as she counted and looked puzzled. I asked her why. She said, "Well, Devon got eight, and I think it's eight too. But when I watched her count, I thought she counted this cube twice so I didn't think she was going to get eight."

Devon didn't seem to follow this line of reasoning, even after I repeated what Jamilla had said. I asked Devon to recount. She did, in a similar manner. I asked Jamilla if she had a strategy that would help her avoid counting any cubes twice. She moved each cube as she counted it and announced, somewhat relieved, "There really were eight."

Counting Teddy Bears Marshall's bag contained teddy bear counters of several colors. He sorted the bears into piles of like colors—two purple, three red, four yellow, and two green—and made tallies on his paper to show each group. Several times he got lost in his counting, and each time this happened he started again at 1. Even though some of the groups were small and all of the bears were visible, he was unable to look at any of the arrangements and count them in his head or recognize a group without counting them.

When I asked Marshall how many bears there were altogether, he went back to the groups of bears and counted them, beginning again several times before announcing that there were 10.

His partner, Emma, heard this and recounted her tally marks. I asked her why she did that. Her response was, "Because I counted 11."

I asked what made her to decide to recount. She told me, "Well, Marshall just said he got 10. And I thought we had 11. We both counted the bears in the bag, so we should have the same number."

Counting Blocks Kenji and Serena had a bag of seven blocks from the block center. Kenji stacked the blocks into a tower and counted them for me. He did not touch the blocks, but pointed to them as he said each number.

Serena tried to mimic Kenji's actions. But she repeatedly got lost in the sequence, looking to us for the name of the number that came next, or guessing the name of a number she'd heard before. ("1, 2, 3 . . . 8?") She would often get further in the sequence if I asked her to start from 1.

Counting Keys Ian was counting keys, which were in a pile on the table. As he counted, he moved the keys from one pile to another. Several times he lost track of where he was, once because he overheard a neighbor's count:

> **Ian**: 5, 6, 7 . . .
> **Neighbor**: 9, 10, 11.
> **Ian**: . . . 12, 13.

When Ian lost track like this, he would push all the keys into one pile and start again from 1. He was eventually able to count the group of nine keys accurately, and I was impressed by his ability to stick with the task.

Observing students as they worked on this activity provided me with a wealth of information about their rote knowledge of the sequence of numbers, their strategies for counting and keeping track of a set of objects, and their comfort with recording their mathematical work. When I realized how much information my observation was providing, I decided to take notes on students' strategies and possible next steps.

Wyatt, Javier, Jamilla, Emma, and Kenji seemed to have a strong grasp of the counting sequence and how it relates to a group of objects. Channary, Devon, and even Marshall and Ian seemed to have a good grasp of the sequence, but needed to develop systems for keeping track of and possibly organizing their counts.

I was somewhat surprised and concerned by Binh and Serena—I realized that Serena needed a lot of practice with the sequence of numbers, and both needed a lot of practice counting small quantities to begin to connect the counting sequence to the set of objects. I made a note to myself to visit them in a small-group format in which they can watch and hear others count and see others' strategies for saying one number for each object. I was also surprised that a handful of students seemed ready to grapple with the issues that arise when different people count the same set of objects and get different numbers.

INVESTIGATION 3

Comparing Towers

Focus Time

Measurement Towers (p. 38)

In an activity that introduces comparing as a way of measuring, students work as a whole group, comparing several classroom objects to a tower of 10 cubes to determine which is longer.

Choice Time

Measuring Table (p. 42)

Students make a measurement standard, a tower of 10 cubes, and compare their tower to each of a collection of objects. They sort the objects into two groups: *Things that are longer than my tower* and *Things that are shorter than my tower.*

Grab and Count: Which Has More? (p. 44)

In this variation on the familiar Grab and Count activity, students grab two handfuls of cubes. They count and build a tower for each handful and then compare the towers to find out which has more cubes. They record the information on paper cube strips.

Compare (p. 46)

Students learn to play Compare, a card game in which players each turn up a number card and compare the two numbers to determine which is larger.

Mathematical Emphasis

- Developing and using language to describe and compare lengths *(longer, shorter, the same)*
- Measuring by direct comparison, i.e., directly comparing two objects to determine which is longer
- Developing strategies for counting and keeping track of quantities
- Representing quantities with objects, pictures, numerals, or words
- Comparing two quantities to find which is more
- Recognizing numerals and connecting them with the quantities they represent

Teacher Support

Teacher Note
Learning About Length (p. 50)

Dialogue Boxes
How Did You Compare? (p. 48)
You Have to Look at the Long Side (p. 51)

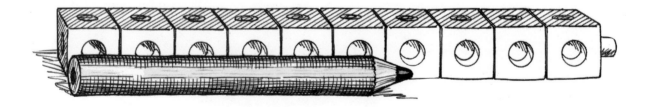

INVESTIGATION 3

What to Plan Ahead of Time

Focus Time Materials

Measurement Towers
- Tower of 10 interlocking cubes
- Collection of objects to measure, such as pens and pencils, glue sticks, rulers, paper towel rolls, bottles, books, boxes, blocks, combs, forks and spoons, small toys. Some should be longer, some shorter than a tower of 10 cubes.

Choice Time Materials

Measuring Table
- Bin of interlocking cubes of one color for each station at the Measuring Table
- Sample 10-cube tower, a *different* color from the cubes in the bins
- Collections of objects to measure, as described for Focus Time, each in its own container

Grab and Count: Which Has More?
- Interlocking cubes separated into bins by color, 10–20 cubes per bin, 2 bins per group
- Crayons to match the colors of the cubes
- Cube Strips (p. 135), with *pairs* of strips cut apart: 3 pairs per student

Compare
- Number Cards, with wild cards and the numbers 7–10 removed: 1 deck per pair. If you do not have manufactured decks, make your own from pp. 126–129.
- Envelopes, resealable plastic bags, or rubber bands for storing the cards
- Interlocking cubes: about 15 per pair

Family Connection
- How to Play Compare (p. 125): 1 per student, for optional homework
- Number Cards (pp. 126–129): 1 set per student, for optional homework

Investigation 3: Comparing Towers

Focus Time

Measurement Towers

What Happens

As a group, students compare the lengths of several objects to the length of a tower of 10 interlocking cubes. They continue making such comparisons on their own during Choice Time, then meet as a group to discuss the strategies they use for comparing lengths. Their work focuses on:

- lining up two objects to determine which is longer
- developing and using language to describe and compare lengths (*longer*, *shorter*, *the same*)

Materials and Preparation

- Snap together a tower of 10 interlocking cubes of one color.
- Provide several collections of objects students can compare to a tower of 10, with some objects shorter, some longer, and some about the same length as 10 cubes. Include some items that have more than one possible dimension to measure and some that are easy to compare by standing them vertically (writing tools, paper towel rolls, bottles). One collection is needed for Focus Time; the rest are for use by students during Choice Time.

Activity

Using Towers to Compare and Measure

This activity introduces comparing as a way of measuring. Display the tower of 10 cubes and the objects to be compared to it.

Sometimes, to see how big something is, we *compare* it with something else. For example, we might stand two people side by side and compare them to see who is taller. Today we're going to use this tower of 10 cubes that way. We will compare the tower to other things to measure them. Who sees something in this collection that looks *longer* than the cube tower?

Ask a volunteer to choose an object from your collection. Encourage students to find the longest dimension of an object when comparing it to the cube tower.

OK, Miyuki, you think this box is longer than my cube tower. Why do you think so? How can we find out for sure? Can you show us how you would do it? ... Does someone have a different way? Can you show us? Does it matter how you hold the box? Does it matter how you hold the tower? Why do you think so?

Investigation 3: Comparing Towers

Ask students to notice how different volunteers compare the tower to the object. Some students will not attend to the longest dimension of an object, and others will hold the tower in a way that does not result in an accurate comparison. You may also find that students verbalize one idea and do something quite different.

Use this opportunity to discuss different methods of comparing the length of two objects so that students are exposed to several strategies they could try. However, do not expect all students to understand or even use these strategies. You will have a chance to discuss these issues in more detail after students have had experience comparing the length of their own tower of 10 to many objects at the Measuring Table activity during Choice Time.

Who sees something in this collection that is *shorter than* my cube tower? . . . Renata says the eraser is shorter. What do we mean when we say it's *shorter* than the tower? How could you compare them to show which is shorter? Does someone have a different way?

After introducing these ideas, explain that during Choice Time students will have many opportunities to compare and measure objects using their own tower of 10. Tell them that everyone needs to work on this activity at some point, as the whole class will be discussing ways they measured and compared the length of different objects.

Focus Time: Measurement Towers ■ **39**

Activity

How Did You Compare?

Note: Students will be comparing lengths in the Choice Time activity, Measuring Table, throughout the rest of this investigation. When everyone has had a chance to work at the Measuring Table, hold several whole-group or small-group discussions in which students can briefly share some of their work.

How Did You Do It? In one type of discussion, ask about the strategies students used for comparing lengths. Have a tower of 10 cubes and a collection of objects so students can model and explain their strategies.

How did you compare your tower to one of these objects? How did you decide whether your tower was longer or shorter than an object?

Ask students to observe and comment on the strategies their classmates used for comparing and to think about whether they used a similar strategy or a different one. See the **Dialogue Box,** How Did You Compare? (p. 48).

Gabriela stood the two objects on the floor and then looked to see which went higher. Did anyone else use that strategy? Would someone compare the tower and the cup a different way? Can you show us how?

When It's Hard to Tell In other discussions, take a close look at objects that students found difficult to place in either the *Longer* or *Shorter* group, or objects that different students placed in different categories.

Some of you put this picture book in the group of things that are longer than the tower. Others put the book in the other group—things that are shorter than the tower. What do you think about that?

Share the item in question with the class, and ask students to demonstrate their methods of comparison. Do they notice that many of these hard-to-place objects have one side that is longer and one side that is shorter than the tower? See the **Dialogue Box,** You Have to Look at the Long Side (p. 51), for one way of handling discussions of this nature.

During these discussions, keep in mind that students learn about comparative measurement by having many opportunities to measure for themselves and to watch others measure. Some students may be starting to understand that comparing in different ways can give different results, while others may not yet be ready to see this. At this point, it is important not to insist that they compare lengths in one particular way. See the **Teacher Note,** Learning About Length (p. 50), for a discussion of how young students come to understand length and ways it can be compared and measured.

Focus Time Follow-Up

A Longer/Shorter Hunt Students go on a hunt to find things in your classroom that are longer than and shorter than a tower of 10. Things that are movable could be placed in two separate spaces, labeled *Longer than my tower* and *Shorter than my tower*. Things that could be placed in either category, depending on the dimension, make good subjects for discussion.

Three Choices In addition to visiting the Measuring Table to continue the activity started during Focus Time, students may also work on two new activities: a variation of Grab and Count called Which Has More? (p. 44), and a card game called Compare (p. 46).

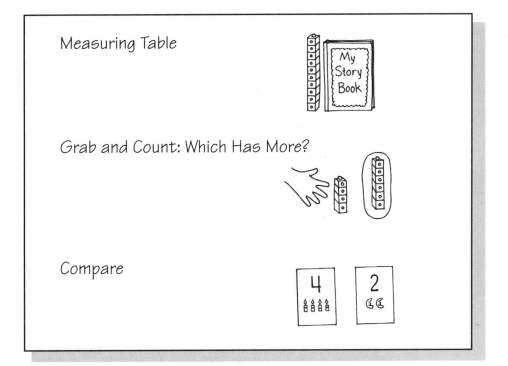

Compare at Home After students have played Compare during Choice Time, you might suggest that they teach the game to someone at home. They will each need a copy of How to Play Compare, a set of Number Cards (0–6 only), and possibly an envelope or plastic bag for storing the cards. Plan to prepare card decks ahead of time for families who may not have scissors at home.

Choice Time

Measuring Table

What Happens

Students now work individually on the activity that was introduced in Focus Time. They each make a cube tower with 10 cubes, then use their towers to sort objects into categories of *Longer than my tower* and *Shorter than my tower*. The class meets periodically as a group to discuss some of the measurement issues students are encountering. Their work focuses on:

- creating a set of a given size
- lining up two objects to determine which is longer
- sorting objects into two categories, according to length

Materials and Preparation

- Set a bin of interlocking cubes at each station on the Measuring Table, one color per bin.
- Provide a model 10-cube tower, a *different* color from the cubes in the bins.
- Provide several collections of materials for measuring, similar to the one used for demonstration in Focus Time, each in its own container.

Activity

Show students the bins of cubes and the model tower at the Measuring Table.

Each of you will need to make yourself a tower with exactly 10 cubes in it, like the model at the table. Count very carefully, and check with the people around you to see if you all agree that everyone has 10.

Show students one of the collections that you have assembled. Briefly demonstrate the activity by sorting a few of the objects with the students.

At the Measuring Table, you will measure each object in a collection and decide whether it is *longer* or *shorter* than your tower of 10. You're going to put all the objects in your collection into two groups: *things longer than my tower* and *things shorter than my tower*.

Remind students that everyone needs to visit the Measuring Table because you will be discussing it as a whole class. The activity How Did You Compare? (p. 40) outlines these discussions. Give students a sense of when the discussions will be held, or wait to schedule them when you see that students have had enough experience with this activity.

Observing the Students

Consider the following as you watch students work at the Measuring Table.

- Do students understand the task? If not, meet with small groups to go over the activity again.

- How comfortable and accurate are students in counting out 10 cubes to make their towers? What kinds of counting errors do you notice? Do students make sure they have 10? If so, how? Do they recount (or ask a classmate to recount) the cubes? Do they compare their tower with a neighbor's? with the model tower? How do they position the towers to compare them?

    If you come across students who do not have 10 cubes in their towers, show them how to compare their tower to the model and how to add or remove cubes as necessary.

- How do students go about sorting each object into one of the two categories? Do they compare the objects directly to their tower? Do they sort some objects without needing to compare them to the tower? How do they sort objects that have one dimension close to the same length as their tower? Can they explain their choice for a given object?

- Are students able to use the vocabulary *longer than* and *shorter than* as they compare objects to their towers?

- How do they position their tower while comparing it to an object? Do they stand items next to each other on the floor or table to compare them? If they lay the tower and object side by side, do they align them at one end?

- What dimensions of objects do students choose to measure? How do they determine which dimension is the longest? Do they ever measure two different dimensions of the same object?

As you observe students, keep track of some of the different methods they use to compare objects to their towers. Also, record any discrepancies that arise, such as an object that some students place in the shorter group and others in the longer group, or an object that seems difficult to place. Use such situations as the focus of your group discussions.

Variations

- Students sort the same objects by comparing them to a cube tower of a different length, such as 5 or 15 cubes. What do students notice about how the groups change?

- Create a third category: *Things about the same size as my tower.* Students might think of the objects in this group as "things that are so close to my cube tower that I can't tell which is longer."

Choice Time

Grab and Count: Which Has More?

What Happens

In this variation of Grab and Count, students grab two handfuls of cubes, count and build a tower for each handful, and compare the towers to determine which of the two quantities is greater. They record this information on paper "cube strips." Their work focuses on:

- counting a set of objects
- comparing two amounts to determine if one is larger
- recording information (data)

Materials and Preparation

- Set out bins of loose cubes of a single color, two bins (two colors) per group.
- Duplicate the Cube Strips (p. 135) and cut in half to make *pairs* of strips. Allow three pairs for each student.
- Provide crayons that match the colors of the cubes.

Activity

This is the first of several variations on Grab and Count, as described in the **Teacher Note**, Grab and Count and Its Variations (p. 91). Gather students around two bins of cubes to introduce this activity.

If I ask Ayesha to grab a handful of these cubes, about how many do you think she will be able to grab? What makes you think so?

If students have been continuing their Choice Time work on Grab and Count, they will likely have a better sense by now of how many cubes they can grab. Ask a volunteer to grab a handful of cubes and place them on the floor. Nearby, set out a pile of five cubes for student reference.

We know this pile has five cubes. About how many cubes do think there are in the handful Ayesha grabbed? Does it look like five? more than five? less than five? Let's count and find out.

Ask a volunteer to count the cubes, or count them together as a class. Snap them together to make a tower.

In this new Grab and Count activity, we're going to compare *two* handfuls. Ayesha grabbed one handful of yellow cubes. Now she's going to grab a handful of a different color.

After your volunteer grabs a second handful, count them and build a second tower. Line up the towers side by side and ask students what they notice. They are likely to observe that the towers are different colors and whether one is longer than the other, although this may not be obvious to everyone. Many kindergartners are just beginning to think about the concepts of longer and shorter, more and less. See what students notice on their own first. If students are struggling, you can prompt them with questions.

Which tower has more cubes? How do you know? Do they have the same number of cubes? Which one has fewer? How do you know?

If your volunteer grabbed the same amount twice, create another example with towers of different heights. If your volunteer grabbed two different amounts the first time, set up and discuss a second example with towers of the same height.

Show the group a pair of cube strips. Gather ideas about how students might color these strips to show the two handfuls of cubes. Students will have different methods for coloring the appropriate number of squares on each strip. Avoid demonstrating any one method so that students will freely choose their own approach.

After students have colored their towers, they circle the tower that has more. If the handfuls are the same, they can circle both towers.

Observing the Students

Consider the following as you watch students working on Grab and Count: Which Has More?

- How do students count their handfuls? Do they know the sequence of number names? Do they count each cube only once? Do they skip or double-count any cubes? Do they organize or arrange the cubes in any way? Do they touch or move each cube as they count it? Do they double-check their count?

- How do students compare their handfuls? What reasoning do they use? Do they compare the towers visually? Do they put them side by side to compare them directly? Do they count the number of cubes in each tower? Can they tell which has more and which has fewer? Do they recognize when the towers are of equal height?

- How do students record their comparisons? Does their work accurately represent their towers?

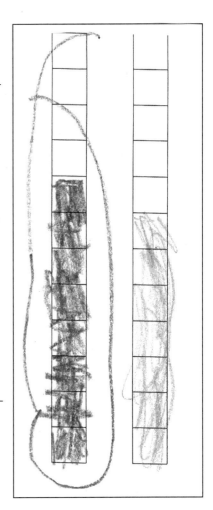

Variation

Students can try the same activity with different materials, such as color tiles, teddy bear counters, or Styrofoam peanuts.

Choice Time: Grab and Count: Which Has More? ■ **45**

Choice Time

Compare

What Happens

In this game, players turn up number cards and determine which card has the larger number. Their work focuses on:

- counting and comparing two numbers or quantities to find which is more
- connecting numerals and number names with the quantities they represent

Materials and Preparation

- Provide decks of Number Cards from 1 to 6 (remove all cards 7–10 and the wild cards).
- Make cubes available, about 15 per pair, for students to use to compare amounts.

Activity

Gather students in a circle on the floor and play a demonstration game of Compare. Choose one student to play with you, or two to play together.

The object of this game is to decide which of two cards shows a larger number. Each player starts with half the cards in the deck.

Demonstrate how to deal out the cards evenly between the two players, using a "one for you, one for me" strategy. Players stack their cards face-down, then both turn up their top card.

Oscar turned over 6, and Shanique turned over 4. Which is larger? How can we figure that out? How do you know?

The person who has the larger number says "me." In this round, who would say "me"? Sometimes you'll both have the same number on your cards. When that happens, you both turn over your next card, and now the person with the higher number says "me."

Explain that the game is over when players have turned over all their cards. Play two or three more rounds, until you think students understand the game. If the zero card has not come up in the demonstration, talk with students about what it means.

Students who have difficulty comparing numbers may want to build a tower of cubes for the number on each card. Players then compare the heights of the towers, and the one with the higher tower says "me."

Observing the Students

Consider the following as you watch students play Compare.

- How do students evenly divide the cards? If they are having difficulty, suggest that one student deal them out and repeat "one for you, one for me" to help keep track.
- Do students understand the rules of the game?
- Do they play cooperatively? If you think students are playing too competitively, emphasize that getting the larger number is a matter of luck, not of being a better player. Explain that good players play cooperatively, for example, by checking or helping one another, by explaining their thinking to one another, by asking one another for help, or by waiting while a partner takes the time to determine which number is larger.
- Can students read and interpret the numerals on the cards, or do they count the objects on the cards to figure out the number? Can they count the objects accurately? If some students are having difficulty distinguishing 6 from 9, remind them to count the objects on the cards to check, and to orient the cards so that the number is on the top.
- What strategies do students have for determining which number is larger? Do they just "know"? Do they count objects on the card? Do they use interlocking cubes?

If some students are struggling, you might call together a small group and work with them as they play. Encourage them to proceed slowly as they find and share ways to count and compare.

Students should play the game Compare many times, both at school and at home. See the **Teacher Note**, Games: The Importance of Playing More Than Once (p. 97). As students grow more familiar with the game, introduce some of the variations.

Variations

- The player who turns up the card with the smaller number says "me."
- Play with three people and compare three numbers for each round. The player who turns up the card with the largest number says "me."
- As students are ready, they can play Compare with card decks that contain all the numbers 0 through 10.

DIALOGUE BOX

How Did You Compare?

This class has been working at the Measuring Table over the past week. Every student has visited the table several times to sort a set of objects into two groups: *Longer than my tower* and *Shorter than my tower*. Students have now gathered to discuss the strategies they have been using to compare lengths.

I'm wondering how you compared different objects to the tower of 10. Did anyone find a way you think works well?

Miyuki: I put them on the table next to each other.

Can you say a little more about that? Let's see, if you were comparing the tray to the tower, you would put them on the table next to each other—like this?

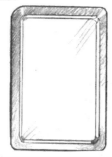

Miyuki: *[rearranging the objects]* No, next to each other the other way. Like this. See? The tray is longer. It goes up and down [beyond the tower at both ends].

Oh, I see. So you put the tower next to the long side of the tray *[traces the long edge of the tray with a finger]*, **and the tray was longer than the tower. Did anyone else do it like Miyuki?** *[Several students raise their hands.]*

Did anyone do it a different way? Would someone compare the tray and the tower a different way?

Tarik: I did like your first way.

Can you show us?

Tarik: Like this. The tower was taller on top and on the bottom.

Interesting! So, this way, the tray is *shorter* than the tower. What do people think about that? Does it matter how you lay the tray?

Tess: Doesn't it need to be straight?

What do people think? Does it matter if the things you are comparing are straight?

Tarik: *[He adjusts the tower so it's straight but does not line up the end with one edge of the tray.]* It doesn't matter. See? The tower's the same—taller on top and on the bottom.

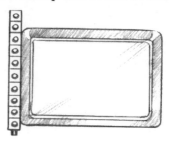

[Other students agree with Tarik that "straightness" doesn't matter.]

Continued on next page

So some of you think it matters if the tower is straight and some of you don't. Let's keep that question in mind as we do more measuring and comparing. We'll continue to think and talk about it together.

Does someone have another way to compare the tower to the tray?

Kylie: I did it sorta the same as Miyuki, but I put the cubes down more, right on the edge of the table. *[She adjusts the arrangement again.]* Then you look to see which goes up more. So it was only taller on top.

So Kylie put them next to each other on the table too, just like Miyuki, but she lined up the bottoms of the two things. Did anyone else use this strategy too? *[Some students raise hands.]* **Any other methods for comparing these two?**

Felipe: I didn't lay them down. I stood them up, like this. Then you can tell the tray is longer because it's taller up. *[He demonstrates but has a hard time standing the tray on its edge.]* This tray doesn't do it so good. It worked good with the bottle and the box though.

Does that way remind anyone of another way we heard about? It reminds me a little bit of Kylie's way. What's the same about them?

This exchange reflects the range of ideas—both accurate and inaccurate—that exist when young students begin to explore measuring. Expect such situations to arise in your classroom, and do not rush to show students how to measure "correctly." Students learn the important ideas in measuring gradually over time and with repeated experiences. You can facilitate their learning by encouraging them to share their strategies and explain their thinking.

Teacher Support ■ **49**

Teacher Note: Learning About Length

In kindergarten and first grade, students start working with ideas about what's *long, longer, short, and shorter.* Their ideas about length begin to develop as they compare lengths directly:

"My sister is taller than I am."

"My pencil is the shortest in the class."

Research on children's mathematical understanding shows that students typically do not develop a firm idea about length as a stable, measurable dimension until toward the end of the primary grades, although of course there is a range of individual differences among students.

Kindergarten students will vary quite a bit in how accurately and consistently they compare the lengths of things. You will probably see some who do not carefully line up objects in order to compare them. You may also see students comparing along a dimension that is not the longest. These "mistakes" are not just carelessness or sloppiness; instead, these students are still figuring out what it means to measure length, and how to go about it.

Rather than simply telling students to align the ends of two objects carefully in order to compare them, or demonstrating how to measure with a tower of cubes, encourage discussion among students about the different ways they are measuring.

Some people thought that this box was longer than our cube tower, and some said it was shorter. Who would like to show how you measured this box? ... Brendan lined up his cube tower like this. Do you think that's OK? Why or why not?

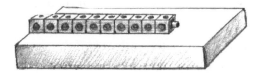

Have fun with this. At times, you might show students some wildly inaccurate ways of measuring in order to help them think through and articulate their own ideas. For example, compare the length of a cube tower with a dimension of an object that is clearly not the longest one.

Or, hold the tower in an unusual way to compare the lengths, such as crosswise horizontally or diagonally.

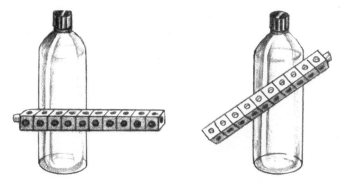

Ask students if each approach is a good way to compare the lengths of the two objects and, if not, what you should do to get a better comparison. As students discuss and compare ways of measuring, they will gradually develop a sense of what length is and how to measure it accurately.

DIALOGUE BOX

You Have to Look at the Long Side

These students have been working at the Measuring Table, comparing lengths of objects to a tower of 10 cubes and placing the objects in groups *longer than my tower* or *shorter than my tower*. Observing that certain objects have been placed differently by different students—sometimes in the *longer* group and sometimes in the *shorter* group—the teacher selects for discussion one such object, a book that is short and wide.

I was at the Measuring Table the other day, and I saw something really interesting. First I saw someone take this book, compare it to the tower, and put it in the group of things that were longer than the tower. Then I saw another person compare this book to a tower and put it in the other group—things that were *shorter* than a tower. Did anyone have any trouble with this book that they'd like to share?

Tiana: Well, I did this book and I found out that it was longer than my tower. But then Shanique did the same book and she said it was shorter. It can't be both. I think I'm right.

What do people think about that? Is that OK? Can an object be in one group one time and in the other group another time?

Thomas: I don't think so, because it's the same thing. It didn't grow or get smaller or change or anything.

Maybe if the two of you could show us how you compared your tower to this book, that would help.

Tiana: I did it like this. See? It's longer than my tower.

Shanique, would you like to show us how you compared the tower and the book? Or, actually, can anyone think of a way that Shanique might have compared this tower and this book and found out that the tower was longer?

Oscar: I think she did it like me. See, Tiana held the book like this. But when you read the book, you hold it like this. So that way, the tower is way longer than the book.

So does it matter which way you hold the book?

Shanique: It matters for when you read it.

Tiana: It matters for reading it. But I think when you measure it, you have to look at the long side and put that by your tower.

So for some objects, there is more than one side that can be measured. I wonder if that's true for every object?

INVESTIGATION 4

Counting and Comparing

Focus Time

Letters in Our Names (p. 54)

Students gain experience with counting and comparing in a context of special interest to them: their own names. After listening to a story about someone with a very long name, students determine how many letters are in that long name and in their own. Everyone builds a name tower with one cube for each letter of his or her name.

Choice Time

Comparing Names (p. 60)

Students use their individual name towers to do some comparing of names in the class. Whose name is longest? Whose is shortest? Whose names are the same length?

Grab and Count: Compare (p. 62)

In another variation of Grab and Count, three or four students in a small group each grab a handful of cubes. They count and build towers for the handfuls. Then they compare handfuls to find which tower has the most cubes, which has the least, and if any are the same.

Collect 10 Together (p. 64)

In this cooperative game, students work in pairs to collect 10 counters. They take turns rolling a dot cube with 1–3 dots to accumulate the counters.

Mathematical Emphasis

- Counting groups of objects
- Creating a set of a given size
- Comparing quantities to determine which is more
- Using terms to describe and compare amounts *(less, least, more, most, same, equal)*
- Keeping track of the size of a growing collection
- Finding the total of two single-digit numbers
- Recording mathematical work

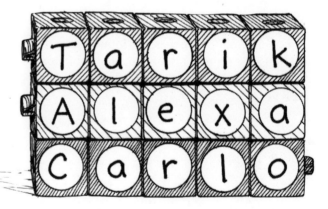

INVESTIGATION 4

What to Plan Ahead of Time

Focus Time Materials

Letters in Our Names
- Name cards: 1 per student
- *Chrysanthemum* by Kevin Henkes (Greenwillow Books, 1991), optional; see p. 54 for alternatives.
- Interlocking cubes, sorted into bins by color
- Dot stickers labeled with ABC's to provide letters for students' name towers

Choice Time Materials

Comparing Names
- Students' name towers from Focus Time
- Student Sheet 3, Our Names (p. 130): 1 per student

Grab and Count: Compare
- Interlocking cubes, sorted into bins by color
- Crayons to match the colors of the cubes
- Cube Strips (p. 135): 3–4 sheets per student, uncut

Collect 10 Together
- Dot cubes with 1–3 dots: 1 per pair
- Counters: 15–20 per pair
- Paper Towers for Collect 10 (p. 132), cut into individual towers; 1 tower per pair, optional

Family Connection
- Investigating Names at Home (p. 140): 1 per student, for optional homework

Investigation 4: Counting and Comparing ■ 53

Focus Time

Letters in Our Names

What Happens

Together the class reads *Chrysanthemum*, a story about a little mouse who loves her name—until she goes to school for the first time. After reading the story, students figure out how many letters are in Chrysanthemum's name and in their own names. They use cubes to make name towers, adding one cube for each letter in their names. Their work focuses on:

- counting a set of objects
- creating a set of a given size

Materials and Preparation

- Obtain a copy of *Chrysanthemum* by Kevin Henkes (Greenwillow Books, 1991) or another book with a long-named character, such as *Tikki Tikki Tembo* by Arlene Mosel (Greenwillow Books, 1968), or *Rumpelstiltskin* by Paul O. Zelinsky (Dutton Children's Press, 1986).
- If you do not have a class set of name cards, print the first name of each student on an unlined 3-by-5-inch card. Decide whether to use full names or the names students are known by.
- Using dot stickers that fit on the sides of your interlocking cubes, label the dots with the letters of the alphabet, capitals and lower case, as needed for students' names. For simplicity, you may want to give each student a strip of dots that spells out his or her name. However, if you want to offer practice in letter recognition, you can provide sheets of dots grouped in rows alphabetically. With this approach, keep in mind that you'll need many more of some letters than of others; you can tally the letters in the names on your class list for an accurate count.
- Prepare one example of a name tower made out of cubes with dot stickers, for your own name or for an absent student.
- Sort interlocking cubes into bins by color.

Activity

Chrysanthemum

Note: This activity uses the book *Chrysanthemum*; it can easily be adapted to a similar story about a character with a long name. If you use the book *Tikki Tikki Tembo*, use only that portion of the name.

Read *Chrysanthemum* aloud to your students. After reading and discussing the book, talk about the length of the mouse's name.

I think everyone agrees that Chrysanthemum is a very long name. How many letters do you think there are in Chrysanthemum? How could we find out exactly how many letters there are in her name?

Encourage students to share different ideas or strategies for finding out how many letters. Some will recall that, in the story, there is mention of 13 letters in the name. While they may not remember how many letters the book said, these students will likely suggest looking through the book for the answer. Others will suggest counting the letters in the name as they appear on the cover or elsewhere in the book. Still others might suggest writing the name down and then counting the letters. Verify the number of letters by testing each strategy that students suggest.

Kylie said to look for the page where it tells us how many letters. We looked there and it said 13. Alexa suggested that we count the letters in the name, on the cover of the book. We did that, and we got 13 again. Now Xing-Qi has suggested writing the name and counting the letters. What if we try that? *[Write* Chrysanthemum *on the board.]* **How many letters do you think there would be if we counted now?**

Some students will realize that no matter how the name looks, it's the same name, with 13 letters. Others may think that since the name written on the board looks so much longer, it must have more than 13 letters. As a group, count the letters in the name again, to double-check.

If you have time and students are interested, you might also count the number of letters for the other long-named character, Delphinium.

Activity

How Many Letters in Your Name?

What if I asked how many letters are in *your* name? How would you figure that out?

Students may suggest a variety of strategies, including writing their names down and counting the letters, finding their names somewhere in the classroom and counting the letters, and spelling their names out loud and counting the letters as they are spoken. After students have shared their ideas, show them the class set of name cards.

Your job today is to figure out how many letters there are in your name. You can use any of the ideas we just talked about for counting the letters. You can also double-check this information using your name card.

Explain to students that once they determine the number of letters in their name, they will build a cube tower with one cube for each letter. Use your name or the name of a child who is absent to demonstrate.

I have a name card here with Luke's name written on it. How many letters are in Luke? If we were going to make a name tower for Luke, how many cubes would we need? How do you know?

As you (or student volunteers) demonstrate, model taking one cube for every letter in the sample name. Then, use the dot stickers to label each cube with a letter. Emphasize that you are using one dot for each cube.

What's the first letter in Luke's name? How do you know? Can someone find me a dot with that letter? I'm going to put the first letter of Luke's name on the first cube. What's the second letter in Luke's name?

Continue until you have a completed name tower. Then explain that students will be doing the same activity with their own names.

The issue of nicknames will likely come up. Encourage students to share their opinions and ideas about this. Explain the decision you made in making the name cards, and change the cards of any students who object to the name on their card.

If you have students who use two names as a first name (such as Rosa Lee or José Luis), you will need to discuss with them how to represent their names as towers. They may decide to leave a blank cube (perhaps with a blank dot sticker) between the two names; they may begin the second name right after the first, marking the change with a new capital letter; or they may use only one of the two names for this activity.

Activity

Making Name Towers

Students spend any remaining time working independently on their name towers. Circulate as students work to find out how they are counting the letters and the cubes.

Observing the Students

Consider the following as you watch students make their name towers.

- How do students go about finding out how many letters are in their names? How do they count the letters? Do they double-check their counts? If so, how? What sorts of counting errors do you notice?

- How do students decide how many cubes to take for their name towers? Do they take one cube for each letter of their name? Do they double-check their counts? How do they double-check?

Instead of taking the same number of cubes as letters in their names, some students may take enough cubes to make a tower that is the same length as their name on the card. For example:

If this happens, ask the student to count the letters on the card and the cubes in the name tower. You might point to the first, then the second, then the third letter, and so on, asking the student to show you the cube for each letter. Alternatively, ask about the student's reasoning, but allow the student to discover that there are not enough (or too many) cubes during the process of putting on the lettered dot stickers.

- How do students label their name towers with lettered dot stickers? Do they use one sticker per cube? What do they do if the number of cubes is not correct?

As students finish, remind them that they will use their name towers to explore the length of everyone's name during an upcoming Choice Time activity. Designate a space or container, such as a tray, for storing name towers until then.

Focus Time Follow-Up

 **Extensions**

First, Middle, and Last Names Students investigate their own first, middle, and last names—how many letters in all? Is your last name longer, shorter, or the same as your first name?

How Many Letters in Our Names? Make a whole-class representation of the number of letters in the names of all the people in your class. Encourage students to think about questions like these: Whose names have the most letters? the fewest letters? Looking at the numbers of letters, which number has the most names? the fewest names?

3	4	5	6	7
Ida	Luke	Jacob	Renata	Justine
	Tess	Alexa	Thomas	Brendan
	Ravi	Maddy	Ayesha	
		Kadim	Felipe	
		Oscar	Miyuki	
		Kylie		
		Tiana		
		Henry		

Homework

Names at Home After comparing names in class, students are often interested in counting the letters in the names of people in their families. Whose name has the most letters? Whose has the fewest? Some may want to put all the names in order by length. Send home Investigating Names at Home (page 139) to provide directions. Plan for some class time when students can share the results of their investigation.

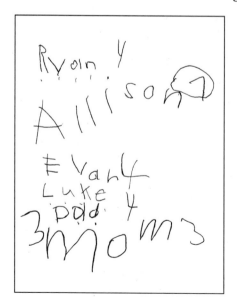

Luke wrote the name of each person in his family. Then he counted the letters in each name and wrote the total next to it. The number 7 by Allison is circled because that's the longest name.

58 ■ *Investigation 4: Counting and Comparing*

Three Choices In addition to Comparing Names, which builds on the work students did during Focus Time, there are two new activities for Choice Time. In Grab and Count: Compare (p. 62), students count several groups of objects and compare amounts. In Collect 10 Together (p. 64), they count and keep track of growing quantities.

Comparing Names

Grab and Count: Compare

Collect 10 Together

Choice Time

Comparing Names

What Happens

Students use the name towers they made during Focus Time to compare the lengths of the names of everyone in the class. Their work focuses on:

- comparing two or more quantities to determine which is greater

Materials and Preparation

- Make available the name towers students made during Focus Time.
- Duplicate Student Sheet 3, Our Names (p. 130), 1 per student.

Activity

| S | h | a | n | i | q | u | e |

| I | d | a |

| H | e | n | r | y |

| T | i | a | n | a |

To introduce this activity, select four name towers from your class: one long, one short, and two medium-length with the same number of letters.

The other day, everyone made a name tower. Today you're going to find out whose name has more letters than yours, whose name has fewer letters, and whose name has the same number of letters.

Hold up one of the medium-length name towers you chose.

Does anyone know whose name tower this is? This is Henry's name. Henry has 1, 2, 3, 4, 5—five letters in his name and five cubes in his tower. Who thinks they see a tower that has more letters than Henry? How can you tell? Does anyone have a different way to tell?

Students may have different strategies for choosing a longer name tower. They might stand the two towers next to each other to see which is longer, count the letters (or cubes) in the towers to see which has more, or count the letters in the names by spelling them out for themselves.

Who thinks they see a name tower on the rug that has fewer letters than Henry's? Justine thinks that the name Ida has fewer letters than Henry. Why do you think so?

Again, encourage students to explain and or demonstrate how they know which name (or tower) has fewer letters (or cubes). Finally, ask students about the last name tower, the one with the same number of letters.

What do you think about Tiana's tower? How could we compare it to Henry's? Do you think it will have more letters? fewer? the same?

Encourage students to share their thoughts and explain their reasoning.

Students will use the name towers to compare their own name with the names of their classmates. Show them Student Sheet 3, Our Names, and demonstrate how to use this chart.

First, at the top, write your name and tell us how many letters are in it. Below, there are three sections—one for names shorter than yours *[point]*, **one in the middle** *[point]* **for names that have the same number of letters as yours, and this last section** *[point]* **for names that are longer than yours. If you find a name with** *fewer* **letters than yours, where do you write it? Where do you write a name with** *more* **letters than yours?**

Remind students that to write each name, they need only copy the letters from the name tower.

Observing the Students

Consider the following as students work on Comparing Names.

- How do students approach this task? Do they compare two towers directly to see which is longer? How do they hold the towers to compare them? Do they line or stand them up? Do they count the letters or cubes in the towers and compare the amounts? Do they sort the towers into piles of names the same length? Are there any names or towers that they place without needing to actually compare them?

- Can they explain why or how they chose where to place a name or tower? What reasoning do they use?

A student with a very long or a very short name may end up recording most other names in just one section. If you have students in this situation, see if they discover this on their own. If they do, they might like to investigate what happens if they compare their nickname or middle name to the first names of their classmates.

Variations

- Find out whose name has the most letters and whose has the least.
- Group everyone who has a name with the same number of letters.
- Order the names of a small group by the number of letters.

Choice Time

Grab and Count: Compare

What Happens

In this variation of Grab and Count, several students grab a handful of cubes, each grabbing cubes of a different color. They count and build towers for each handful. Then they compare handfuls by asking: Which tower had the most cubes? the least? Were any the same? Their work focuses on:

- comparing quantities to determine which has more
- using language to describe and compare amounts *(least, less, most, more, same, equal)*

Materials and Preparation

- Sort loose cubes by color into three or four bins.
- Provide crayons to match the colors of the cubes.
- Duplicate Cube Strips (p. 135), 3–4 sheets per student. For this activity, use whole (uncut) sheets.

Activity

Remind students of their work with Grab and Count (p. 12) and the variation Grab and Count: Which Has More? (p. 44).

What do we know about the size of a handful of cubes? If I ask Thomas to grab a handful of these cubes, about how many do you think he will be able to grab? What if I ask Tess? How many do you think she can grab? more than Thomas? less? the same? Why do you think so?

Today we're going to do another variation of the activity Grab and Count. This time, we're going to compare different people's handfuls. The Grab and Count station can hold three or four people working together. Each person will grab a handful of cubes, count how many in that handful, and build a tower with the cubes.

Do a demonstration round, having three or four volunteers each grab a handful of cubes of a different color, count them, and build a tower. Stand the towers next to each other, not in any particular order.

What can you tell me about these four towers?

Encourage students to refer to the towers by color, saying, for example, "The red one has the most," instead of, "Thomas grabbed the most." This can help avoid feelings of competition or self-consciousness.

If necessary, prompt the students with questions. If students are having trouble comparing three or four objects at once, you might ask them to talk about two towers, and then add the other towers as students are ready.

Do some towers have more than others? Which ones? Which tower has the most? Which tower or towers have the least? Do any of the towers have the same number of cubes?

For all responses, ask students to explain or show how they know. Demonstrate how to record on the Cube Strips. Students color in each handful on one strip, and then circle the one with the most.

Renata said the tower with nine had the most. What if two people both grabbed nine cubes? How would we record that?

As the students begin the activity on their own, remind them that all the students at the Grab and Count station together will work together, making towers to compare.

You'll be thinking about these questions: Which tower (or color) has the most? Which has the least? Do any have the same?

Observing the Students

Consider the following as you watch students working on Grab and Count: Compare.

- How do students count their handfuls? Do they know the sequence of number names? Do they count each item once and only once? Do they organize the cubes in any way? Do they double-check their count?

- How do students compare their handfuls? What reasoning do they use? Do they compare the towers by looking at them? by setting them next to each other? Do they count the number of cubes in each tower? Can they tell which has more (or the most)? which has less (or the least)? Do they recognize when towers are equal? How comfortable are students comparing more than two handfuls? more than three?

- How do small groups of students work together? Do they work cooperatively? Do they double-check the counts? How do they handle disagreements?

    If you find students are becoming too competitive, remind them to compare towers by color rather than as "yours" and "mine." Explain that you are not interested in who can grab the most, but rather in students' strategies for comparing several towers.

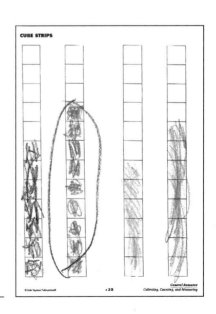

Choice Time

Collect 10 Together

What Happens

Students play the game Collect 10 Together, rolling a dot cube to accumulate counters until they have 10. Their work focuses on:

- finding the number of dots in dot patterns
- counting a group of objects
- keeping track of the size of a growing collection of objects
- finding the total of two single-digit numbers

Materials and Preparation

- Make dot cubes with 1–3 dots using stickers on blank 1-inch cubes. Cover two sides each with one dot, two dots, and three dots.
- Provide 15–20 counters, such as buttons or bread tabs, for each pair.
- Duplicate Paper Towers for Collect 10 (p. 132) and cut apart; provide an individual tower to any pair who wants to use one to keep track.

Activity

Introduce Collect 10 Together by playing a demonstration game. Play one or two complete games to be sure students understand how to play.

The goal of Collect 10 Together is for two players to collect 10 counters, working together. You take turns rolling a dot cube. Then you take as many counters as there are dots on the cube.

Ask a volunteer to inspect the dot cube and tell the class about it. When you have established that there are 1, 2, and 3 dots on the sides of the cube, ask first one player and then the other to roll the die.

What did Kadim roll? How many counters does he take for the set? How do you know? . . . What did Tiana roll? How many counters does she add to the set? How do you know?

After each player has rolled and everyone has agreed on how many counters to take, talk about the growing set.

Do you think Kadim and Tiana have 10 counters yet? Why do you think so? How many do they have? How do you know?

64 ■ *Investigation 4: Counting and Comparing*

Ask two or three students to explain how they could figure out how many counters there are in all. Some students may count from 1. Others might count on: "There were 2, and 2 more makes 3, 4." Others may use knowledge of addition pairs: "I know that 2 and 2 makes 4." Encourage students to explain why they think the total number is more or less than 10.

Where in our classroom could we find out if 4 is more or less than 10? In what other ways could we compare the numbers 4 and 10?

Introduce the Paper Towers for Collect 10 as a tool for keeping track of the accumulated counters. The game is over as soon as players have at least 10 counters. Depending on how many dots they roll at the end, they may get 10 exactly, or they may end up with more than 10.

Observing the Students

Consider the following as you watch students play Collect 10 Together.

- Do students understand how to play the game? Do they play cooperatively? Do they double-check each other's counting?

- Do students instantly recognize the number of dots on the cube, or do they count them to see how many counters to take?

- Do students take the same number of counters as dots on the cube?

- How comfortable and accurate are students as they count? What sorts of errors do you notice in their counting? Are they able to keep track of what has been counted and what needs to be counted?

- How do students find the total number of counters after the new ones have been added? Do they count all the counters from 1? count on from the number they had after the last turn? Do they "just know" some combinations that are 1 (or 2, or 3) more?

- Do they recognize when they have 10 counters? Do they recognize when they have more than 10? Do they use a paper tower to keep track?

If students are having difficulty comparing amounts to 10, you might change the game to Collect 7 or 8 Together. Students ready for more challenge might collect a greater number, such as 12 or 15.

Variation

Use Number Cards for 1's, 2's, and 3's instead of dot cubes. Players shuffle the deck and turn it facedown. On each turn, a player turns over the top card and collects that many counters.

INVESTIGATION 5

Least to Most

Focus Time

Least to Most (p. 68)

In this investigation, students extend their experience with comparing amounts as they order three or four towers of cubes by length or by amount. In the introductory activity, students watch a volunteer grab several handfuls of cubes. For each handful, students help count the cubes and build a tower. They then compare and order the towers from least to most. Finally, they create a representation of the towers to show the results.

Choice Time

Grab and Count: Least to Most (p. 72)

Students work on their own to grab several handfuls of cubes and build a tower for each handful. They put the towers in order from least to most and make representations of their work.

Racing Bears (p. 74)

In this simple board game, students roll a dot cube and advance four teddy bear counters along separate tracks of 10. When a bear gets to the 10 space, the players collect a counter.

Continuing from Investigation 4
Collect 10 Together (p. 64)

Mathematical Emphasis

- Counting sets of objects
- Comparing quantities to determine which is more
- Learning and using appropriate language to describe and compare amounts (*least, less, most, more, same, equal*)
- Ordering quantities from least to most or most to least
- Representing mathematical work

Teacher Support

Dialogue Box
Comparing and Ordering Towers (p. 76)

INVESTIGATION 5

What to Plan Ahead of Time

Focus Time Materials

Least to Most
- Interlocking cubes, sorted into bins by color
- Crayons that match the cubes
- Cube Strips (p. 135): 1 sheet, cut into single strips
- Unlined paper: 1 sheet
- Glue stick or tape

Choice Time Materials

Grab and Count: Least to Most
- Cubes and crayons from Focus Time
- Cube Strips (p. 135): 1–2 sheets per student, cut into single strips
- Unlined paper: 1–2 sheets per student
- Glue sticks or tape

Collect 10 Together
- Dot cubes with 1–3 dots: 1 per pair
- Counters such as buttons or bread tabs: 15–20 per pair
- Paper Towers for Collect 10 (p. 132) from Investigation 4

Racing Bears
- Racing Bears Gameboard (p. 133): 1 per pair
- Teddy bear counters: 4 per pair
- Small counters such as buttons or bread tabs: about 10 per pair
- Dot cubes with 1–3 dots (and with 1–6 dots for more challenge): 1 cube per pair of players

Family Connection
- How to Play Collect 10 Together (p. 131): 1 per student for optional homework. Also send home Number Cards for the numbers 1–3 (pp. 126–127) if students don't already have a complete set at home.

Investigation 5: Least to Most

Focus Time

Least to Most

What Happens

Focus Time introduces the idea of ordering several cube towers by size. Together the class gathers four handfuls of cubes, counts them, builds four towers, and then orders the towers from least to most. Paper cube strips are used to record the results. Students' work focuses on:

- counting sets of objects
- comparing several quantities to determine which is more
- using appropriate language to describe and compare amounts *(less, least, more, most, same, equal)*
- ordering quantities from least to most or most to least
- using a recording sheet to represent mathematical work

Materials and Preparation

- Sort interlocking cubes into bins by color. You need at least four bins.
- Provide crayons that match the colors of the cubes.
- Duplicate a sheet of Cube Strips (p. 135) and cut apart the four strips.
- Have at hand unlined paper and a glue stick or tape.

Activity

Four Handfuls: Least to Most

Note: The group demonstration of this activity uses four handfuls of cubes. You can adjust for different levels of challenge by suggesting different numbers of handfuls as students work on this activity during Choice Time.

Ask four volunteers to each grab, count, and build a tower for a handful of cubes, each taking cubes of a different color. Using different colors helps eliminate confusion when students are discussing the towers.

Line up the four towers next to each other, but not in order by length. Ask students what they notice about the towers. If students are struggling to respond, you can prompt them with questions.

Which tower has the most cubes? How do you know? Do any have the same number of cubes? Are any equal? Which tower has the least cubes? How do you know?

68 ▪ *Investigation 5: Least to Most*

After everyone has had a chance to look carefully at the towers and make observations, ask the class to help you order the towers.

What does it mean to put a group of things in order? If I wanted to put these towers in order, how might I do that? Are there different ways we could order these towers? What's one way we could do it? . . . Does someone have a different way?

What if I wanted to put these towers in order from the smallest handful to the biggest handful? How might I do it? What would you do first?

Give students a chance to share and demonstrate their ideas. Some students might suggest starting with the tower with the most or least cubes, while others might group same-size towers together and work from there.

Demonstrate how students will record their work. Show them the blank cube strips and explain that this task is similar to recording students have already done with other Grab and Count activities. This time, they will record each handful on a single strip, put the paper strips in order, and then tape or glue them to a sheet of unlined paper. Quickly demonstrate the process for the four cube towers you have been discussing.

Explain that during Choice Time, students will grab, count, and order their own handfuls of cubes. It is important for every student to do this activity because the whole class will be sharing their strategies for counting, comparing, and ordering towers of cubes.

Activity

Talking About Least to Most

Note: Students will be working on Grab and Count: Least to Most over several days of Choice Time. After everyone has spent some time on the activity, hold several discussions to share students' work. These discussions might focus on strategies for counting, strategies for comparing and ordering, and considering typical handfuls.

Strategies for Counting Have a container of cubes available for students to use as they explain their counting strategies. Encourage students to share their strategies for keeping track of what has been counted and what still needs to be counted.

When you grabbed a handful of cubes, how did you count them? What strategies did you use to find out how many there were? How did you keep track?

Ask the class to comment on the different strategies that are described and to think about whether they used a similar strategy or a different one.

Alexa snapped the cubes into a tower and counted each cube as she put it on. Did anyone else use that strategy? Did someone count their cubes a different way? How?

While some students may begin to grasp their classmates' strategies, others will still be working to develop one strategy of their own. While kindergartners are beginning to make sense of numbers and counting, it is important not to insist that they count in one particular way. Refer to the **Teacher Notes,** Counting Is More Than 1, 2, 3 (p. 16) and Observing Students As They Count (p. 32), for more information.

Strategies for Comparing and Ordering In another type of discussion, focus on the strategies students used to compare and order their towers. Again, have cubes and perhaps some sample towers for students to use as they explain their strategies and reasoning. See the **Dialogue Box,** Comparing and Ordering Towers (p. 76), for a typical discussion.

How did you compare your handfuls to put them in order?

Can you show me which tower has the most cubes? How do you know? the least? How do you know? Do any have the same? How do you know?

Students need a lot of experience, both watching and doing, to build their ordering skills. It is important to let them develop their own approaches and strategies as they are learning.

Typical Handfuls After the class has had a lot of experience with this activity, consider holding a discussion about a "typical" handful.

Miyuki grabbed five different handfuls. *[Point to each tower as you mention it.]* **Once she grabbed 4, three times she grabbed 6, and one time she grabbed 7. What can we say about how many cubes Miyuki usually grabs? If I ask Miyuki to take another handful now, how many do you predict she will grab? Why do you think so?**

Do you think Miyuki would *ever* grab 14 cubes? 18? 25? Why do you think so?

Focus Time Follow-Up

Three Choices Students continue the activity Grab and Count: Least to Most, which was introduced during Focus Time. From time to time, bring the group together to discuss the counting and comparing issues that come up. Students also begin a new activity, the game Racing Bears (p. 74), and continue the activity Collect 10 Together from Investigation 4 (p. 64).

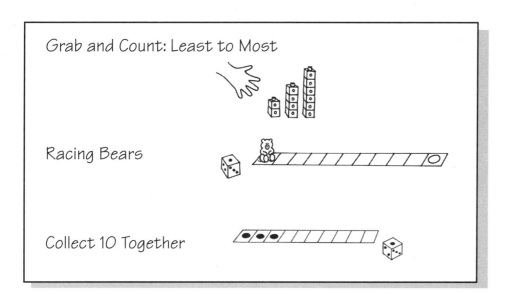

Collect 10 Together at Home As students continue playing Collect 10 Together during Choice Time, you might also ask them to play the game with someone at home. They will need a copy of How to Play Collect 10 Together (p. 131), and a set of twelve Number Cards, four each of the numbers 1, 2, and 3 (pp. 126–127).

Choice Time

Grab and Count: Least to Most

What Happens

Students work individually on the activity that was introduced in Focus Time. They grab several handfuls of cubes, count them, build towers with them, and put the towers in order by size. Their work focuses on:

- comparing several quantities to determine which is more
- using appropriate language to describe and compare amounts *(less, least, more, most, same, equal)*
- ordering quantities from least to most or most to least
- using a recording sheet to represent mathematical work

Materials and Preparation

- Provide interlocking cubes, sorted into bins by color.
- Duplicate Cube Strips (p. 135), making 1–2 sheets per student. Cut apart the individual strips.
- Provide crayons that match the colors of the cubes, 1–2 sheets of unlined paper for each student, and glue sticks or tape to share.

Activity

Students work on their own to grab handfuls of cubes, count and build a tower for each handful, and then compare and order all the towers. Adjust the level of challenge by asking students to work with different numbers of handfuls (3 for less challenge, 5 to 7 for more challenge). Remind students to grab a different color for each handful.

Once you have all your towers, you'll put them in order, just the way we did together, from the handful that had the least cubes to the handful that had the most cubes.

Students will record what they find out by coloring in a cube strip for each handful and then gluing them in order on a sheet of paper.

If you notice some students having a hard time comparing and ordering handfuls, ask questions to help them find a way to think about the task.

Can you show me which tower has the most (or least) cubes?

What if we compared just these two towers to start? What can you tell me about these two? OK, the red one has more than the yellow one. I see that you put the yellow one first, and then the red one. What if we added the green one now? Where might that one go?

Another option is for such students to work with a smaller number of handfuls, or to continue working on previous Grab and Count activities until they seem ready for this one.

While students are working on this Choice Time activity, hold whole-group discussions from time to time so students can share what they are finding out. See Talking About Least to Most (p. 70) for discussion ideas.

Observing the Students

Consider the following as students work on Grab and Count: Least to Most.

- How do students count their handfuls? Do they organize the cubes in any way? Do they count each cube once and only once? Do they have a system for keeping track of what's been counted and what still remains to be counted? Do they double-check their count? How comfortable are they with the sequence of number names?

- How do students describe and compare their towers? Do they use comparative terms such as *most* and *least?* Can they identify two towers as being the *same* or *equal*?

- How do students order the towers? Do they begin with the tower that has the most or least cubes? Do they begin by grouping same-size towers? Do they compare pairs of towers? Do they order them visually, building staircases with them? Do they order them numerically, using the number of cubes in each tower?

- How do students represent their results? Does their work accurately represent the handfuls they grabbed?

Variations

- After ordering a set of towers, students grab another handful, build the tower, and place it so that all of the towers are still in order. How do they go about placing that tower?
- What if we grabbed handfuls of something bigger than cubes, like thread spools? Would we be able to grab more or less? What if we grabbed something smaller than cubes, like buttons? Would we be able to grab more or less? Students first predict, then investigate by actually grabbing and counting handfuls of those materials.

Choice Time

Racing Bears

What Happens

To play Racing Bears, students roll a dot cube and move one or more of four bears the indicated number of spaces along a track. The object is to land each bear on the tenth space, to collect a counter. Students' work focuses on:

- finding the number of dots in familiar dot patterns in order to move on a gameboard
- counting out amounts up to six
- becoming familiar with combinations of numbers up to about six

Materials and Preparation

- Use blank 1-inch cubes and stick-on labels to make several dot cubes with 1–3 dots and, for players ready for more challenge, dot cubes with 1–6 dots.
- Provide each pair of players with a copy of the Racing Bears Gameboard (p. 133), a dot cube (either 1–3 or 1–6 dots), four teddy bear counters, and about 10 other small counters such as buttons or pennies.

Activity

Racing Bears can be played individually or by pairs or small groups playing cooperatively. Show students how to set up the gameboard, placing a teddy bear counter at the beginning of each of the four tracks and a counter such as a button in the circle at the end of each track.

Players take turns rolling a dot cube and moving any of the bears on the board that number of spaces. The object is to work together to get a bear to the tenth space on any track. When a bear lands exactly on the tenth space, the players take the counter and replace it with another. Then they move the bear back to the starting space on that track.

Let's play a round together. The first player rolls the dot cube.... What number did Brendan roll? How many spaces can he move? How do you know? Brendan, you can choose which bear to move.

Now it's my turn.... How many spaces can I move? How do you know?

As it becomes appropriate in the demonstration game, ask students to decide which bear to move in order to land on a counter.

A player can use part of a roll to move a bear to the tenth space and then use the rest of the roll to move another bear on another track. For example, suppose a player rolls a 3. The red bear is on the eighth space of its track. The player moves the red bear two spaces to get a counter and puts the bear back on the starting space. Then, to use the rest of the roll of 3, the player moves the green bear one space.

If this situation doesn't arise naturally, use one of your turns to demonstrate what happens when you roll a number that would take a particular bear *past* the tenth space.

I rolled a 3, but I want to move the blue bear, which is two spaces from the end. *[Move the blue bear the two spaces.]* **I rolled 3, and I moved 2. How many moves do I have left? How do you know?**

You may need to demonstrate this type of move several times before students get used to the idea of splitting a roll. If you find that students are having difficulty with split rolls, you might adjust the rules to say that when a bear reaches the tenth space, that is the end of a turn, whether or not there are leftovers from the roll.

Play several rounds, asking students to explain how they figured out how many spaces to move and how they decided which bear to move. The game is over when the players together have collected at least 10 counters.

Observing the Students

Consider the following as you watch students play Racing Bears.

- Do students understand the rules of the game? Do they play cooperatively? Do they double-check each other's moves?

- Do students instantly recognize the number of dots on the cube, or do they count the dots to see how many spaces to move? Do they move the same number of spaces as there are dots on the cube?

- How comfortable and accurate are students as they move and count? What sorts of errors do you notice in their counting?

- How do students choose which bear to move? Do they review their options to see if any of the bears can land on 10? Do they know (or can they figure out) how many more spaces a bear has to go to land on 10?

- How do students handle rolls that would take a bear past 10? Do they keep track of how many spaces they have moved (or counted) and how many they have yet to move? How do they keep track? Do they count the dots? count on their fingers? use number combinations?

DIALOGUE BOX

Comparing and Ordering Towers

This teacher is doing the Focus Time activity: Least to Most with the class. A volunteer has grabbed four handfuls of cubes and built four towers.

Now we have four towers in four different colors. I'm going to line them up next to each other. What do you notice? What can you tell me about these towers?

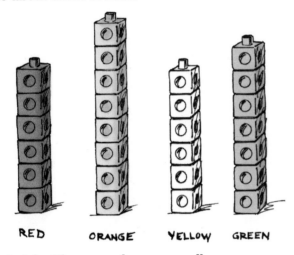

Gabriela: They go red, orange, yellow, green.

Jacob: They look like a rainbow.

Can you compare them? Do some towers have more than others?

Kadim: The green one has more than the yellow one. See how it goes up one higher?

Oscar: Red and yellow have the same, 1, 2, 3, 4, 5, 6.

So she grabbed six cubes twice.

Jacob: Red is the smallest.

Justine: Well, red's tied with yellow, so they're both the smallest. *[She places the red tower next to the yellow to compare them, then counts the cubes in each tower to check.]* Yup. Six is the smallest.

So the smallest handfuls had six cubes. What else do people notice?

Henry: Orange is the most. It's the tallest.

Kadim: Yeah. It's two taller.

Can you show us what you mean?

Kadim: *[Puts the orange and red towers next to each other]* The red is two steps away from the orange.

Henry: Two dots away *[referring to the circles on the cube faces]*.

What if we wanted to put these towers in order? Does anyone have an idea about how to do that?

Ravi: I know! You, um . . . um . . . What do you mean, *in order*?

What if we wanted to line them up so that they went from smallest to biggest? How might you do that?

Alexa: Just look at them. Make them like stairs.

Can you show us what you mean?

Alexa: Well, this one is short and so is yellow. They're a tie. Like a bottom step of the stairs. Then green is a little bit bigger and orange wins. *[She arranges cubes in the order shown.]*

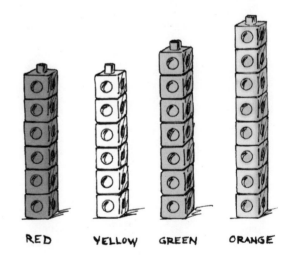

Investigation 5: Least to Most

So Alexa looked at the towers and thought about stairs. Can everyone see how these remind Alexa of stairs? Would anybody put the towers in order in a different way?

Brendan: You could trade the yellow and red . . . *[He does this and sits back.]* That's a different way.

Interesting. Brendan switched the towers that were the same. Are the towers still in order? *[The class agrees.]* **Why can he do that?**

Alexa: Because they're the same.

Brendan: Yeah, they're both the smallest. They both have six so it doesn't matter.

[Mixes the four towers again.] **Would anyone put the towers in order a different way? How would you start?**

Shanique: Well, I'd put the same ones together.

The same ones . . . Can you show me which ones you mean?

Shanique: This one is the same as this one *[she holds the yellow up to the red, then lays them together off to the side.]* And this one is the same as *[holds the orange up to the green]* . . . nope. They don't match. Orange is bigger.

I see now what you mean by putting "same ones" together. What would you do next, to put them in order?

Shanique: These are the smallest. This one has *[silently counts the cubes in the yellow tower]* . . . six. *[Next she counts the red tower.]* So does this one. *[She lays red next to the yellow.]* Green is . . . *[she counts silently]* seven, so it goes next, and the big one has *[counts to herself again]* . . . eight.

So Shanique put towers next to each other. She put ones that were the same together, and then I noticed that she counted how many cubes were in the towers to help her put them in order. Does anyone have a different way to put these towers in order?

When this discussion began, students were describing data, making comparisons, and talking about the ideas *more than, less than,* and *equal to.* They used these relationships later as they discussed how to put the cubes in order, a new concept for many of the students.

INVESTIGATION 6

Arrangements of Six

Focus Time

Six Tiles (p. 80)

Students find many different ways to arrange six tiles so that adjacent tiles touch in some way. They share arrangements and talk about how they might remember different arrangements of six.

Choice Time

Books of Six (p. 86)

Students continue to explore arrangements of six. They share their favorites and then combine all the different arrangements they have discovered into Books of Six.

Continuing from Investigation 5

Collect 10 Together (p. 64)

Grab and Count: Least to Most (p. 72)

Racing Bears (p. 74)

Mathematical Emphasis

- Finding different ways to visualize and arrange a set of six objects
- Solving a problem with many possible solutions
- Developing strategies for counting and keeping track of quantities through about 12
- Representing quantities with objects, pictures, or numerals
- Comparing quantities to determine which is more
- Developing and using language to describe and compare amounts *(less, least, more, most, same, equal)*
- Ordering from least to most

Teacher Support

Teacher Notes

Why Six? (p. 88)

From the Classroom: "I Had One Too Much" (p. 89)

Grab and Count and Its Variations (p. 91)

Dialogue Box

It Looks Like a Chair (p. 90)

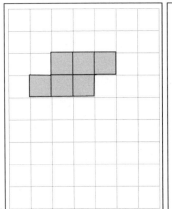 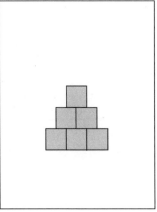 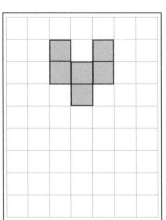 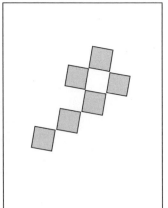

INVESTIGATION 6

What to Plan Ahead of Time

Focus Time Materials

Six Tiles

- Color tiles: at least 6 per student, more if available
- Construction paper squares in colors to match the tiles: 30–40 per student
- One-inch grid paper (p. 134) and a supply of unlined paper: 5–8 sheets per student (includes supply for Choice Time)
- Glue sticks

Choice Time Materials

Books of Six

- Color tiles, one-inch grid paper, unlined paper, paper squares, and glue sticks rom Focus Time
- A stapler or a hole punch and string or ribbon for binding students' books (optional)

Collect 10 Together

- Dot cubes with 1–3 dots: 1 per pair
- Counters such as buttons or bread tabs: 15–20 per pair
- Paper Towers for Collect 10 (p. 132): available for use as needed

Grab and Count: Least to Most

- Cubes, separated into bins by color
- Cube Strips, cut into single strips: supply remaining from previous Choice Time
- Crayons, unlined paper, and glue sticks or tape

Racing Bears

- Racing Bears Gameboard: 1 per pair
- Teddy bear counters: 4 per pair
- Small counters such as buttons or bread tabs: about 10 per pair
- Dot cubes with 1–3 dots, or with 1–6 dots for more challenge: 1 per pair

Focus Time

Six Tiles

What Happens

Students use color tiles to make many different arrangements of six with each tile touching at least one other tile. Each student chooses one arrangement to record. Students share their different arrangements and talk about how they might name or remember them. ("Mine looks like stairs." "Mine looks like a T.") Their work focuses on:

- exploring different ways to visualize and arrange a set of six objects
- recording arrangements of objects
- solving a problem with many possible solutions

Materials and Preparation

- Make available at least 6 color tiles per student, more if available.
- Provide one-inch grid paper (p. 134) and unlined paper, 5–8 sheets per student (1–2 sheets for Focus Time, and the rest for Choice Time).
- Cut one-inch squares from construction paper in colors to match the tiles. Either use a paper cutter or copy one-inch grid paper onto the colored paper and cut apart the squares. Prepare about 30–40 squares per student (a sheet of 9-by-12-inch paper yields 108 one-inch squares).
- Set out containers of glue.

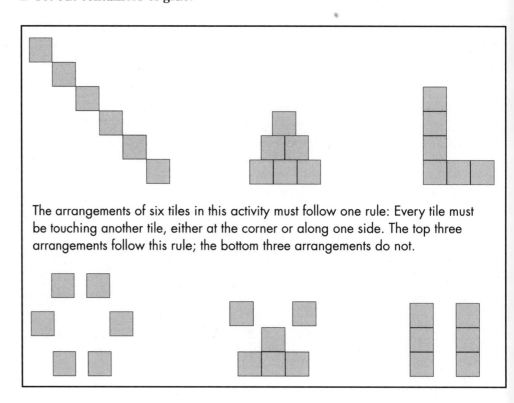

The arrangements of six tiles in this activity must follow one rule: Every tile must be touching another tile, either at the corner or along one side. The top three arrangements follow this rule; the bottom three arrangements do not.

80 ■ *Investigation 6: Arrangements of Six*

Activity

Arranging Six Tiles

Note: For information on why the number six is the focus of this activity, see the **Teacher Note**, Why Six? (p. 88).

With the class, count out five tiles as a reference. Then set out a pile of six tiles and ask students to use their growing skills of estimation to say about how many tiles are in the pile.

Observing the Students

Notice the strategies students use and listen to their reasoning to get a sense of how they estimate and count small groups.

- Do students count (or try to count) the tiles from where they are sitting? Do any students seem to have instant recognition of a group of six?

- What information or reasoning do students use to estimate how many tiles there are? Do they use the reference pile of five? Do they think back to previous experience with grabbing handfuls of tiles? Do they use visual information? ("I know it can't be 10. It doesn't look like that many.") Do they use number combinations? ("I can see four blues and two reds, and 4 and 2 more is 6.")

When students have shared their predictions and explained their ideas, ask one or several students to count the tiles, reinforcing the idea of double-checking as an important strategy when counting.

Today we're going to use tiles to make arrangements of six. When I put out these tiles, I just scattered them here on the floor. Does someone have an idea of how we might *arrange* these tiles to show six? Who has an idea about what the word *arrange* means?

After students have shared their ideas, explain that in this activity there is one special rule about what makes an arrangement of six tiles: The tiles all need to *touch* in some way. They can touch at the corners, share part of a side, or share an entire side. Arrangements do *not* follow the rule if any tile or group of tiles is not touching another tile. Make several sample arrangements to show what you mean, including some that do *not* follow the rule. (See examples on page 80.)

Ask students to think for a moment about how they might arrange six tiles. Then ask a volunteer to show one idea to the class.

Here is Tarik's way. How many tiles do you think there are in his arrangement? Do you agree that there are six? How do you know? Does Tarik's way follow the rule? Do all the tiles touch in some way?

Is there another way we could arrange these tiles to show six? Does someone have a different idea?

Give several students a chance to share their ideas. Observe to see if students count the number of tiles each time or if they are secure with the fact that there are still six since none was added or taken away. If you have students who do not count the tiles, ask them to explain why they do not. Alternatively, consider using a new set of six tiles each time so that you can leave each arrangement in place for comparison.

Once students understand the task and the fact that there is more than one possible arrangement, introduce the following activity.

Activity

My Six Tiles

Your job today is to find different ways to arrange six tiles. Remember that all the tiles need to touch. You can make as many different arrangements as you like, but everyone will need to choose one way to record. Then we will come together and share them.

Show students the grid paper, the unlined paper, and the paper squares. Collect ideas about how students might use these materials to record.

For the next 20 minutes or so, students work individually making arrangements of six tiles. Circulate to be sure they understand the task and to observe how they are counting, arranging, visualizing, and recording their arrangements. As you talk with students, encourage them to explain how they know they have six, and to think of ways to name or describe their arrangements. You might make notes on students' papers as they name or describe the arrangement they chose to record.

Students will view the arrangements in a variety of ways. Strategies might focus on the overall visual image ("It looks like a staircase"), the colors ("It goes red, blue, blue, yellow, yellow, yellow"), the number of tiles in different parts of the arrangement ("It goes 1 then 2 then 3"), or some combination of these ("It's like steps—1, 2, 3"). Some students may not have anything to say except "It's six." While it's important to encourage students to think of grouping the tiles mentally in some way, it may be sufficiently challenging for some to just count out and record six things. Others may have created an arrangement that does not in fact look like anything to them.

82 ■ *Investigation 6: Arrangements of Six*

For students who are ready to think about this question but can't think of a way to describe their arrangements, ask others at the table to share their approaches. Or, choose a part of a student's arrangement, name it, and ask the student to finish naming the arrangement. For example:

You have four up here in a group. How many are in the bottom group?

How many blue tiles did you use? [5 blue] **How many red?** [1 red] **So one way we could think about yours is "five blue tiles and one red tile." Could you think of another way?**

Tess said hers looks like a T. Does yours look like something to you?

Observing the Students

Consider the following as you watch students arranging six tiles.

- Do students make arrangements of six and only six tiles? Does each tile in their arrangements touch another tile in some way?

- Can students accurately count out six tiles? Do they double-check their count on their own? at your request? Can they look and tell that a pile "doesn't look like six"? If they don't have six, do they adjust the number of tiles? How? If there are five tiles, do they add one more? If there are seven, do they take one away? Do they count the tiles from 1 every time?

- How do students arrange the tiles? Do the arrangements seem random? Can students explain how they arranged the tiles? Do they find ways that are easy to remember or explain visually? numerically? both?

- Do students realize there are many possible solutions to the problem? How comfortable are they with this? Do they find more than one arrangement on their own? at your suggestion?

- How do students think about and describe their arrangements? What kinds of strategies do they have for remembering them? Are their strategies primarily visual? color-related? numerical? a combination of these?

- How do students record their work? Do they use grid paper or unlined paper? Do they build on the paper and then remove one tile at a time to glue in a paper square? Do they build their arrangement next to the paper and refer back and forth? Do they count the tiles in groups to transfer the information? ("The first row has 3 . . . 1, 2, 3. Now, the second row has 2 . . . 1, 2.") How do they keep track? Do their representations match their tile arrangements?

As students work, they may acquire an extra tile or misplace one of the six they are working with. This might be a counting issue, or it may be the result of several children working in close proximity. Encourage students to double-check their counting and to tell you how they are sure they have six tiles. See the **Teacher Note**, From the Classroom: "I Had One Too Much" (p. 89), for one teacher's observations while watching students work on this activity.

Activity

Sharing Our Arrangements

Ask students to bring the one arrangement they recorded to share with the group. Begin by encouraging students to talk about what (if any) strategies they used to make sure they always had six tiles.

How did you make sure you were always working with six tiles? What did you do if you found out you didn't have six tiles?

This is a good opportunity to talk about double-checking as an important strategy in counting. Ask students to share what they did if they counted and did not have six tiles.

What if these were your tiles? There aren't six in this pile. What would you do?

Spend some time sharing the actual arrangements students made. Ask students to hold up their representations, and, if they like, tell about them. There will be a variety of responses. Some students will attend to the color or number of tiles; some will consider their arrangement just "6" or "a design" or "a thing"; others will see the overall image of the tiles as an object or letter.

After students have shared their arrangements, choose one design with an obvious shape (such as a T) and ask students how they would remember it if you asked them to build it. Accept different ways of viewing and remembering the arrangement. The **Dialogue Box**, It Looks Like a Chair (p. 90), illustrates how students in one class saw a particular arrangement.

Focus Time Follow-Up

Sorting Arrangements of Six Use the students' recordings of arrangements of six tiles to set up a new Choice Time activity. Students sort the collected recordings into groups that they think go together.

3-D Arrangements of Six Students might be interested in a similar activity using materials that include 3-D possibilities, such cubes. That is, how many arrangements of six interlocking cubes can they make?

Four Choices For the upcoming Choice Time, students continue to investigate and make arrangements of six tiles that they then combine into individual Books of Six. In addition, they may continue to work on three familiar counting activities from earlier investigations: Collect 10 Together (p. 64), Grab and Count: Least to Most (p. 72), and Racing Bears (p. 74).

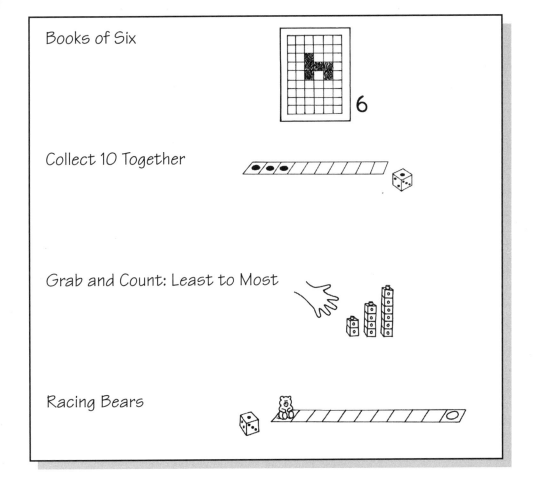

Choice Time

Books of Six

What Happens

Students continue to find different ways to generate and record arrangements of six color tiles. They record arrangements and combine their pages to make their own individual Book of Six. Their work focuses on:

- finding different ways to visualize and arrange a set of six objects
- recording arrangements of objects
- solving a problem with many possible solutions

Materials and Preparation

- Make available the color tiles, sheets of grid paper and unlined paper, colored paper squares, and glue sticks from Focus Time.
- If you plan to bind students' books, have available a stapler or a hole-punch and string or ribbon.

Activity

Ask students to review the work they've already done arranging six tiles. Remind students of the rule that adjacent tiles need to touch in some way.

We're going to continue to find different ways to show six with tiles. Everyone has already recorded at least one way. Now you're going to see how many different ways you can find. Record each new arrangement on a sheet of plain paper or grid paper. Then, each of you will make your own Book of Six by putting your sheets together.

As necessary, review strategies for recording arrangements. Explain where students will store their work (perhaps in a folder or cubby) until they are ready to combine all their recordings into a book.

While students are working on this Choice Time activity, hold whole-group discussions from time to time so students can share their arrangements and their strategies for recording them. Whenever you see an interesting new arrangement, put it on display. Ask students how they would describe it, or how they would remember the arrangement if you asked them to build it with tiles.

When students have recorded as many arrangements of six tiles as they can, help them combine their pages into a book. Staple the pages together, or use a hole punch and string or ribbon to bind them. Students might like to design a cover for their book, perhaps illustrated with their favorite way to make six.

Observing the Students

Consider the following as you watch students working on their Books of Six.

- Do students make arrangements of six and only six tiles? Does each tile in their arrangements touch another tile in some way?

- Can students accurately count out six tiles? Do they double-check to make sure they have six? If they don't have six, do they adjust the number of tiles? How? Do they count them from 1 every time? If there are five tiles, do they add one more? If there are seven, do they remove one?

- How do students arrange the tiles? Do they organize them in any way? What kinds of strategies do they have for describing or remembering them? Are their strategies primarily visual? color-related? numerical? a combination of these?

- How do students record their work? Do they build on the paper and then remove one tile at a time to glue in a paper square? Do they build their arrangement next to the paper and refer back and forth? Do they count the tiles in groups to transfer the information? How do they keep track? Do their representations match their tile arrangements?

Variation

Make a display of the different arrangements of six the class has found. As students find more ways, they can check the class set and see if each idea is in fact new. If it is, then it can be added to the display. Eventually, you could use this display to make a whole-class book of ways to make six.

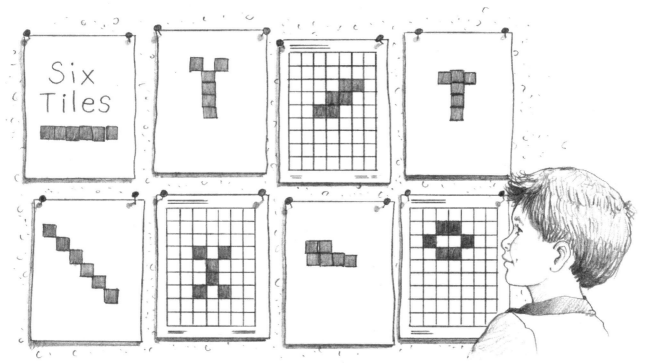

Teacher Note: Why Six?

Through the activities Six Tiles and Books of Six, students explore the quantity six. While the same activities could have been done with almost any other single-digit number, six was chosen for some very specific reasons. Our experience indicates that six is an amount most kindergartners can count with some accuracy, even early in the year. Because students need more than one hand to represent six on their fingers, they will naturally work with combinations of two numbers. This gives them lots of opportunities to represent and see six as combinations of other numbers—1 and 5, 2 and 4, 3 and 3, 2 and 2 and 2, and so forth. In addition, six is one of the largest amounts that can be mentally visualized, mentally manipulated, and instantly recalled. Finally—and most important to kindergarten students—many of them are likely to be either 6 years old or turning 6 during the year.

The focus on six also supports students' work later in *Investigations*. The counting and comparing activities in this unit are crucial to building an understanding of quantities up to six that will enable students to think about and build combinations of numbers that make six and quantities larger than six.

It may seem appropriate to have students work with every number in this way, so they get repeated practice counting a given number of objects and seeing those objects arranged in a variety of ways. However, when asked to do the same activity with every number, students soon grow bored with the activity, and the mathematical experience is lost.

When kindergartners work with six tiles, some will have primarily visual ways of seeing the arrangements, while others may see them in numerical parts.

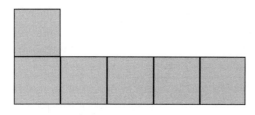

"It's a laying down L."
"It looks like my brother's hockey stick."
"It's 1 and 5."
"I see 2 on the end and then 4."

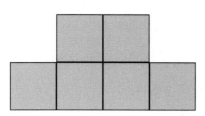

"It looks like a hat."
"It's a thing you step up on."
"It's 3 this way and 3 that way."
"It goes 1 and 2 and 2 and 1."
"I see 2 on top and 4 under it."

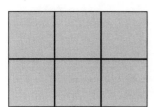

"It's a box."
"It's a window with 6 pieces of glass."
"I remember 3 and 3."
"It's 2 and 2 and 2."

FROM THE CLASSROOM

"I Had One Too Much"

Teacher Note

I had just introduced Choice Time, with the activity Books of Six as one of the choices. While students worked to generate their different arrangements of six, I visited that station to observe their strategies for counting, arranging, and recording their work with the tiles.

Immediately I noticed that several of the students were not working with six tiles. One boy had fewer than six, and three others all clearly had more. Every available seat at this station was taken, and tiles were scattered around the table.

I began by asking the students how many tiles they were using. Everyone responded, "Six." The question prompted one student to check her own count, which was six, as well as her neighbor's, which was eight. When she pointed this out, her neighbor objected, then counted and realized she had two extra tiles. She separated out six tiles to work with.

Since none of the other students checked their counts, I asked them how they were sure they had six and suggested counting to double-check. The boy was surprised to find that he had fewer than six. He looked around for an extra, eyeing the pile of the student next to him. He didn't see any strays, so he took another tile from the bin and began making a new arrangement. I was surprised he didn't feel the need to count again, and asked him how he knew he needed one more. He replied, "Because we're supposed to have six. I only had five."

Then I watched another student try to count his tiles. Because they were randomly scattered, he had a hard time—he couldn't keep track of which ones he'd already counted. He touched several tiles more than once, but seemed to know something was up when he hit numbers above ten. Finally he pushed all the tiles back to the center of the table and slowly and carefully counted out a new group of six.

One girl double-checked her count by putting all her tiles together and moving each tile to one side as she counted it. When she finished counting (at seven), she took one tile and put it back in the center of the table. She told me, "I had one too much." Then she recounted her tiles, and, satisfied, began making new arrangements with them.

As I continued to watch students work, I noticed on several occasions that a tile or two shifted from one student's pile to another's. There were also several tiles on the floor. I suggested that students keep all spare tiles in the container, to lessen the confusion, but I also decided that this station was too crowded. It was too hard for children to organize and keep track of their work. The next time students do this activity, I will limit the number working on it.

I often have mixed feelings about limiting the number of students at an activity, especially when there seems to be a lot of interest in it. As I plan, I always consider the available space and materials, but when the choice is introduced and more kids want to go to that activity, I usually give in. I need to try assuring my students that an activity will be available for a long time and that everyone who is interested will be able to participate. Space and materials really do matter and often have an impact on the success or failure of an activity and a student's experience with it.

Teacher Support

DIALOGUE BOX

It Looks Like a Chair

This class has been working on the Choice Time activity, Books of Six, for several days. Students have had the time to generate and record several different arrangements of six tiles. Now they are meeting to share some of their work. After each student shares a design, the teacher chooses one, an L shape, to talk about how students would remember it.

Take a minute to look at and think about Miyuki's arrangement. What do you notice?

Luke: It has greens and yellows.

Tiana: Yeah, and there's a pattern *[pointing]*: green, yellow, green, yellow, green, yellow.

So Luke and Tiana noticed the colors in Miyuki's pattern, and how the colors were arranged. How else would you remember this pattern?

Brendan: It looks like a L.

Luke: There's six. *[Counts the tiles one by one]* 1, 2, 3, 4, 5, 6.

Tess: I think it looks like a chair.

Ravi: Or a leg and a foot.

Luke: My name starts with L!

Tess: Well, there's three up and down *[holds up three fingers on one hand]*, and then there's three across on the bottom *[holds up three fingers on the other hand]*. That's six.

So how do you know there are six?

Tess *[touching each finger to her chin as she counts]*: 1, 2, 3, 4, 5, 6.

Did you see what Tess used to figure that out? Yes, she used her fingers. I've noticed lots of kids using their fingers to count and keep track.

Grab and Count and Its Variations

Teacher Note

Grab and Count is an activity that kindergarten students never seem to tire of. The basic activity and its variations offer repeated experience with counting, comparing, organizing, keeping track of, and representing quantities. This is a good example of how teachers can take a basic activity, and, once it is familiar to students, change it in a variety of ways to build on students' previous experiences while also extending the mathematics they are investigating. Because the structure is familiar to them, students can revisit the activity fairly easily and independently. In any of the Grab and Count activities, you can vary the size of the objects to be grabbed and thereby adjust the numbers that students will be working with, to offer different students an appropriate level of challenge.

Grab and Count In the basic version of Grab and Count (Investigation 1, p. 12), students practice counting reasonable quantities of a variety of materials. After grabbing a handful of cubes, color tiles, teddy bear counters, or other objects, they count to find out how many they were able to grab. Students find their own ways to represent the information about their handfuls on paper (Student Sheet 1).

In the basic Grab and Count activity, this student grabbed a handful of Styrofoam peanuts. He traced each one and then counted them, drawing a line under each picture to help him keep track. He then recorded the number 6.

Grab and Count: Which Has More? This variation is introduced as a Choice Time activity in Investigation 3 (p. 44). Now students grab two handfuls of cubes, each of a different color. They count the cubes, build towers, and compare them, thinking about which handful had more cubes and how they know. Finally, students use a sheet with two "cube strips" to represent their handfuls. On the paper, they circle the strip with more.

Students will take different approaches to the recording process. Some may place a cube tower adjacent to the paper cube strip and color corresponding squares. Others may lay the tower directly on the strip and remove one cube at a time to color the square beneath it. Still others may count the cubes in a tower and then count and color the same number of squares on the strip. None of these approaches is better than the others, and students should be encouraged to find their own recording method.

Grab and Count: Compare In this variation introduced for Choice Time in Investigation 4 (p. 62), three or four students compare several towers made from handfuls of cubes and think about these questions: Which tower had the most? the least? Did any have the same? How do you know? Using a full sheet of cube strips, they color a strip of squares for each tower and circle the one(s) with the most.

Grab and Count: Least to Most This version is introduced in Investigation 5 (p. 68). Again, students compare several handfuls of cubes, and now also order the quantities from least to most. After building cube towers and putting them in order, students represent the information by coloring in individual cube strips and ordering those strips on a sheet of paper.

Teacher Note: About Choice Time

Choice Time is an opportunity for students to work on a variety of activities that focus on similar mathematical content. In the kindergarten *Investigations* curriculum, Choice Time is a regular feature that follows each whole-group Focus Time. The activities in Choice Time are not sequential; as students move among them, they continually revisit the important concepts and ideas they are learning in that unit. Many Choice Time activities are designed with the intent that students will work on them more than once. As they play a game a second or third time, use a material over and over again, or solve many similar problems, students are able to refine their strategies, see a variety of approaches, and bring new knowledge to familiar experiences.

Scheduling Choice Time

Scheduling of the suggested Choice Time activities will depend on the structure of your classroom day. Many kindergarten teachers already have some type of "activity time" built into their daily schedule, and the Choice Time activities described in each investigation can easily be presented during these times. Some classrooms have a designated math time once a day or at least three or four times a week. In these cases you might spend one or two math times on a Focus Time activity, followed by five to seven days of Choice Time during math, with students choosing among three or four activities. New activities can be added every few days.

Setting Up the Choices

Many kindergarten teachers set up the Choice Time activities at centers or stations around the room. At each center, students will find the materials needed to complete the activity. Other teachers prefer to keep materials stored in a central location; students then take the materials they need to a designated workplace. In either case, materials should be readily accessible. When choosing an arrangement, you may need to experiment with a few different structures before finding the setup that works best for you and your students.

We suggest that you limit the number of students doing a Choice Time activity at any one time. In many cases, the quantity of materials available establishes the limit. Even if this is not the case, limiting the number is advisable because it gives students the opportunity to work in smaller groups. It also gives them a chance to do some choices more than once.

In the quantity of materials specified for each Choice Time activity, "per pair" refers to the number of students who will be doing that activity at the same time (usually not the entire class). You can plan the actual quantity needed for your class once you decide how many other activities will be available at the same time.

Many kindergarten teachers use some form of chart or Choice Board that tells which activities are available and for how many students. This organizer can be as simple as a list of the activities on chart paper, each activity identified with a little sketch. Ideas for pictures to help identify each different activity are found with the blackline masters for each kindergarten unit.

In some classrooms, teachers make permanent Choice Boards by attaching small hooks or Velcro strips onto a large board or heavy cardboard. The choices are written on individual strips and hung on the board. Next to each choice are additional hooks or Velcro pieces that indicate the number of students who can be working at that activity. Students each have a small name tag that they are responsible for moving around the Choice Board as they proceed from activity to activity.

Introducing New Choices

Choice Time activities are suggested at the end of each Focus Time. Plan to introduce these gradually, over a few days, rather than all at once on the same day. Often two or three of the choices will be familiar to students already, either because they are a direct extension of the Focus Time activity or because they are continuing from a previous investigation. On the first day of

Choice Time, you might begin with the familiar activities and perhaps introduce one new activity. On subsequent days, one or two new activities can be introduced to students as you get them started on their Choice Time work. Most teachers find it both more efficient and more effective to introduce activities to the whole class at once.

Managing Choice Time

During the first weeks of Choice Time, you will need to take an active role in helping students learn the routine, your expectations, and how to plan what they do. We do not recommend organizing students into groups and circulating the groups every 15–20 minutes. For some students, this set time may be too long to spend at an activity; others may have only begun to explore the activity when it's time to move on. Instead, we recommend that you support students in making their own decisions about the activities they do. Making choices, planning their time, and taking responsibility for their own learning are important aspects of the school experience. If some students return to the same activity over and over again without trying other choices, suggest that they make a different first choice and then do the favorite activity as a second choice.

When a new choice is introduced, many students want to do it first. Initially you will need to give lots of reassurance that every student will have the chance to try each choice.

As students become more familiar with the Choice Time routine and the classroom structure, they will come to trust that activities are available for many days at a time.

For some activities, students will have a "product" to save or share. Some teachers provide folders where students can keep their work for each unit. Other teachers collect students' work in a central spot, then file it in individual student folders. In kindergarten, many of the products will not be on tidy sheets of paper. Instead, students will be making constructions out of pattern blocks and interlocking cubes, drawing graphs on large pieces of drawing paper, and creating patterns on long strips of paper.

Continued on next page

Teacher Note continued

For some activities, such as the counting games they play again and again, there may be no actual "product." For this reason, some teachers take photographs or jot down short anecdotal observations to record the work of their kindergarten students.

During the second half of the year, or when students seem very comfortable with Choice Time, you might consider asking them to keep track of the choices they have completed. This can be set up in one of these ways:

- Students each have a blank sheet of paper. When they have completed an activity, they record its name or picture on the paper.

- Post a sheet of lined paper at each station, or a sheet for each choice at the front of the room. At the top of the sheet, write the name of one activity with the corresponding picture. When students have completed an activity, they print their name on the appropriate sheet.

Some teachers keep a date stamp at each station or at the front of the room, making it easy for students to record the date as well. As they complete each choice, students place in a designated spot any work they have done during that activity.

In addition to learning about how to make choices and how to work productively on their own, students should be expected to take responsibility for cleaning up and returning materials to their appropriate storage locations. This requires a certain amount of organization on the part of the teacher—making sure storage bins are clearly labeled, and offering some instruction about how to clean up and how to care for the various materials. Giving students a "5 minutes until cleanup" warning before the end of any Choice Time session allows students to finish what they are working on and prepare for the upcoming transition.

At the end of a Choice Time, spend a few minutes discussing with students what went smoothly, what sorts of issues arose and how they were resolved, and what students enjoyed or found difficult. Encourage students to be involved in the process of finding solutions to problems that come up in the classroom. In doing so, they take some responsibility for their own behavior and become involved with establishing classroom policies.

Observing and Working with Students

During the initial weeks of Choice Time, much of your time will be spent in classroom management, circulating around the room, helping students get settled into activities, and monitoring the process of making choices and moving from one activity to another. Once routines are familiar and well established, however, students will become more independent and responsible for their own work. At this point, you will have time to observe and listen to students while they work. You might plan to meet with individual students, pairs, or small groups that need help; you might focus on students you haven't had a chance to observe before; or you might do individual assessments. The section About Assessment (p. I-8) explains the importance of this type of observation in the kindergarten curriculum and offers some suggestions for recording and using your observations.

Materials as Tools for Learning

Teacher Note

Concrete materials are used throughout the *Investigations* curriculum as tools for learning. Students of all ages benefit from being able to use materials to model problems and explain their thinking.

The more available materials are, the more likely students are to use them. Having materials available means that they are readily accessible and that students are allowed to make decisions about which tools to use and when to use them. In much the same way that you choose the best tool to use for certain projects or tasks, students also should be encouraged to think about which material best meets their needs. To store manipulatives where they are easily accessible to the class, many teachers use plastic tubs or shoe boxes arranged on a bookshelf or along a windowsill. This storage can hold pattern blocks, Geoblocks, interlocking cubes, square tiles, counters such as buttons or bread tabs, and paper for student use.

It is important to encourage all students to use materials. If manipulatives are used only when someone is having difficulty, students can get the mistaken idea that using materials is a less sophisticated and less valued way of solving a problem. Encourage students to talk about how they used certain materials. They should see how different people, including the teacher, use a variety of materials in solving the same problem.

Introducing a New Material: Free Exploration
Students need time to explore a new material before using it in structured activities. By freely exploring a material, students will discover many of its important characteristics and will have some understanding of when it might make sense to use it. Although some free exploration should be done during regular math time, many teachers make materials available to students during free times or before or after school.

Each new material may present particular issues that you will want to discuss with your students. For example, to head off the natural tendency of some children to make guns with the interlocking cubes, you might establish a rule of "no weapons in the classroom." Some students like to build very tall structures with the Geoblocks. You may want to specify certain places where tall structures can be made—for example, on the floor in a particular corner—so that when they come crashing down, they are contained in that area.

Establishing Routines for Using Materials
Establish clear expectations about how materials will be used and cared for. Consider asking the students to suggest rules for how materials should and should not be used; they are often more attentive to rules and policies that they have helped create.

Initially you may need to place buckets of materials close to students as they work. Gradually, students should be expected to decide what they need and get materials on their own.

Plan a cleanup routine at the end of each class. Making an announcement a few minutes before the end of a work period helps prepare students for the transition that is about to occur. You can then give students several minutes to return materials to their containers and double-check the floor for any stray materials. Most teachers find that establishing routines for using and caring for materials at the beginning of the year is well worth the time and effort.

Teacher Note

Encouraging Students to Think, Reason, and Share Ideas

Students need to take an active role in mathematics class. They must do more than get correct answers; they must think critically about their ideas, give reasons for their answers, and communicate their ideas to others. Reflecting on one's thinking and learning is a challenge for all learners, but even the youngest students can begin to engage in this important aspect of mathematics learning.

Teachers can help students develop their thinking and reasoning. By asking "How did you find your answer?" or "How do you know?" you encourage students to explain their thinking. If these questions evoke answers such as "I just knew it" or no response at all, you might reflect back something you observed as they were working, such as "I noticed that you made two towers of cubes when you were solving this problem." This gives students a concrete example they can use in thinking about and explaining how they found their solutions.

You can also encourage students to record their ideas by building concrete models, drawing pictures, or starting to print numbers and words. Just as we encourage students to draw pictures that tell stories before they are fluent readers and writers, we should help them see that their mathematical ideas can be recorded on paper. When students are called on to share this work with the class, they learn that their mathematical thinking is valued and they develop confidence in their ideas. Because communicating about ideas is central to learning mathematics, it is important to establish the expectation that students will describe their work and their thinking, even in kindergarten.

There is a delicate balance between the value of having students share their thinking and the ability of 5- and 6-year-olds to sit and listen for extended periods of time. In kindergarten classrooms where we observed the best discussions, talking about mathematical ideas and sharing work from a math activity were as much as part of the classroom culture as sitting together to listen to a story, to talk about a new activity, or to anticipate an upcoming event.

Early in the school year, whole-class discussions are best kept short and focused. For example, after exploring pattern blocks, students might simply share experiences with the new material in a discussion structured almost as list-making:

What did you notice about pattern blocks? Who can tell us something different?

With questions like these, lots of students can participate without one student taking a lot of time.

Later in the year, when students are sharing their strategies for solving problems, you can use questions that allow many students to participate at once by raising their hands. For example:

Luke just shared that he solved the problem by counting out one cube for every person in our classroom. Who else solved the problem the same way Luke did?

In this way, you acknowledge the work of many students without everyone sharing individually.

Sometimes all students should have a chance to share their math work. You might set up a special "sharing shelf" or display area to set out or post student work. By gathering the class around the shelf or display, you can easily discuss the work of every student.

The ability to reflect on one's own thinking and to consider the ideas of others evolves over time, but even young students can begin to understand that an important part of doing mathematics is being able to explain your ideas and give reasons for your answers. In the process, they see that there can be many ways of finding solutions to the same problem. Over the year, your students will become more comfortable thinking about their solution methods, explaining them to others, and listening to their classmates explain theirs.

Games: The Importance of Playing More Than Once

Teacher Note

Games are used throughout the *Investigations* curriculum as a vehicle for engaging students in important mathematical ideas. The game format is one that most students enjoy, so the potential for repeated experiences with a concept or skill is great. Because most games involve at least one other player, students are likely to learn strategies from each other whether they are playing cooperatively or competitively.

The more times students play a mathematical game, the more opportunities they have to practice important skills and to think and reason mathematically. The first time or two that students play, they focus on learning the rules. Once they have mastered the rules, their real work with the mathematical content begins.

For example, when students play the card game Compare, they practice counting and comparing two quantities up to 10. As they continue to play over days and weeks, they become familiar with the numerals to 10 and the quantities they represent. Later in the year, they build on this knowledge as they play Double Compare, a similar game in which they add and compare quantities up to 12. For many students, repeated experience with these two games leads them quite naturally to reasoning about numbers and number combinations, and to exploring relationships among number combinations.

Similarly, a number of games in *Pattern Trains and Hopscotch Paths* build and reinforce students' experience with repeating patterns. As students play Make a Train, Break the Train, and Add On, they construct and extend a variety of repeating patterns and are led to consider the idea that these linear patterns are constructed of units that repeat over and over again.

Games in the geometry unit *Making Shapes and Building Blocks,* such as Geoblock Match-Up, Build a Block, and Fill the Hexagons, expose students again and again to the structure of shapes and ways that shapes can be combined to make other shapes.

Students need many opportunities to play mathematical games, not just during math time, but other times as well: in the early morning as students arrive, during indoor recess, or as choices when other work is finished. Games played as homework can be a wonderful way of communicating with parents. Do not feel limited to those times when games are specifically suggested as homework in the curriculum; some teachers send home games even more frequently. One teacher made up "game packs" for loan, placing directions and needed materials in resealable plastic bags, and used these as homework assignments throughout the year. Students often checked out game packs to take home, even on days when homework was not assigned.

About Classroom Routines

Attendance

Taking the daily attendance and talking about who is and who is not in school are familiar activities in many kindergarten classrooms. Through the Attendance routine, students get repeated practice in counting a quantity that is significant to them: the number of people in their class. This is real data that they see, work with, and relate to every day. As they count the boys and girls in their class or the cubes in the attendance stick, they are counting quantities into the 20s. They begin to see the need to develop strategies for counting, including ways to double-check and to organize or keep track of a count.

Counting is an important mathematical idea in the kindergarten curriculum. As students count, they are learning how our number system is constructed, and they are building the knowledge they need to begin to solve numerical problems. They are also developing critical understandings about how numbers are related to each other and how the counting sequence is related to the quantities they are counting.

In *Investigations,* students are introduced to the Attendance routine during the first unit of the kindergarten sequence, *Mathematical Thinking in Kindergarten.* The basic activity is described here, followed by suggested variations for daily use throughout the school year.

The Attendance routine, with its many variations, is a powerful activity for 5- and 6-year-olds and one they never seem to tire of, perhaps because it deals with a topic that is of high interest: themselves and their classmates!

Materials and Preparation

The Attendance routine involves an attendance stick and name cards or "name pins" to be used with a display board. (Many teachers begin the year with name cards and later substitute name pins as a tool for recording the data.)

To make the attendance stick you need interlocking cubes of a single color, one for each class member, and dot stickers to number the cubes.

To make name cards, print each student's first name on a small card (about 2 by 3 inches). Add a photo if possible. If you don't have school photos or camera and film, you might ask students to bring in small photos of themselves from home.

For "name pins," print each student's name on both sides of a clothespin, being sure the name is right side up whether the clip is to the right or to the left.

Name cards might be displayed in two rows on the floor or on a display board. The board should have "Here" and "Not Here" sections, each divided into as many rows or columns as there are students in your class. To display name cards on the board, you might use pockets, cup hooks, or small pieces of Velcro or magnetic tape. Name pins can be clipped down the sides of a sturdy vertical board.

Collecting Attendance Data

How Many Are We? With the whole group, establish the total number of students in the class this year by going around the circle and counting the number of children present.

Encourage students to count aloud with you. The power of the group can often get the class as a whole much further in the counting sequence than many individuals could actually count. While one or two children may be able to count to the total number of students in the class, do not be surprised or concerned if, by the end of your count, you are the lone voice. Students learn the counting sequence and how to count by having many opportunities to count, and to see and hear others counting.

When you have counted those present, acknowledge any absent students and add them to the total number in your class.

Appendix: About Classroom Routines ■ **99**

About Classroom Routines

Counting Around the Circle Counting Around the Circle is a way to count and double-check the number of students in a group. Designate one person in the circle as the first person and begin counting off. That is, the first person says "1," the second person says "2," and so on around the circle. As students are learning how to count around the circle, you can help by pointing to the person whose turn it is to count. Some students will likely need help with identifying the next number in the counting sequence. Encouraging students to help each other figure out what number might come next establishes a climate of asking for and giving help to others.

Counting Around the Circle takes some time for students to grasp—both the procedure itself and its meaning. For some students, it will not be apparent that the number they say stands for the number of people who have counted thus far. A common response from kindergartners first learning to count off is to relate the number they say to a very familiar number, their age. Expect someone to say, for example, "I'm not 8, I'm 5!" Be prepared to explain that the purpose of counting off is to find out how many students are in the circle, and that the number 8 stands for the people who have been counted so far.

Representing Attendance Data

The Attendance Stick An attendance stick is a concrete model, made from interlocking cubes, that represents the total number of students in the classroom. For young students, part of knowing that there are 25 students in the class is seeing a representation of 25 students. The purpose of this classroom routine is not only to familiarize students with the counting sequence of numbers above 10, but also to help students relate these numbers to the quantities that they represent.

To make an attendance stick, distribute an interlocking cube to each student in the class. After counting the number of students present, turn their attention to the cubes.

We just figured out that there are [25] students in our classroom today. When you came to group meeting this morning, I gave everybody one cube. Suppose we collected all the cubes and snapped them together. How many cubes do you think we would have?

Collect each student's cube and snap them together into a vertical tower or stick. Encourage students to count with you as you add on cubes. Also add cubes for any absent students.

Ayesha is not here today. Right now our stick has 24 cubes in it because there are 24 students in school today. If we add Ayesha's cube, how many cubes will be in our stick?

Using small dot stickers, number the cubes. Display the attendance stick prominently in the group meeting area and refer to it each time you take attendance.

By counting around, we found that 22 of you are here today. Let's count up to 22 on the attendance stick. Count with me: 1, 2, 3 . . . *[when you reach 22, snap off the remaining cubes].* **So this is how many students are not here—who wants to count them?**

In this way, every day the class sees the attendance stick divided into two parts to represent the students HERE and NOT HERE.

Name Cards or Name Pins Name cards or pins are another concrete way to represent the students. Whereas the attendance stick represents *how many students* are in the class, name cards or pins provide additional data about *who* these people are.

Once students can recognize their name in print, they can simply select their card or pin from the class collection as they enter the classroom each day. At a group meeting, the names can be displayed to show who is here and who is not here, perhaps as a graph on the floor or on some type of display board.

Examining Attendance Data

Comparing Groups In addition to counting, the Attendance routine offers experience with part-whole relationships as students divide the total number into groups, such as PRESENT and ABSENT (HERE and NOT HERE) or GIRLS and BOYS. As they compare these groups, they are beginning to analyze the data and compare quantities: Which is more? Which is less? *How many* more or less? While the numbers for the groups can change on any given day, the sum of the two groups remains the same. Understanding part-whole relationships is a central part of both sound number sense and a facility with numbers.

The attendance stick and the name cards or name pins are useful tools for representing and comparing groups. One day you might use the attendance stick to count and compare how many students are present and absent; another day you might use name cards or pins the same way. Once students are familiar with the routine, you can represent the same data using more than one tool.

To compare groups, choose a day when everyone is in school. Count the number of boys and the number of girls.

Are there more boys than girls? How do you know? How many more?

Have the boys make a line and the girls make a line opposite them. Count the number of students in each line and compare the two lines.

Which has more? How many more?

Use the name cards or the attendance stick to double-check this information.

Once the total number of boys and girls is established, you can use this information to make daily comparisons.

Count the number of girls. Are all the girls HERE today? If not, how many are NOT HERE? How do you know? Can we show this information using the name cards? *[Repeat for the boys.]*

If we know that two girls and two boys are NOT HERE, how many in all are NOT HERE in school today? How do you know? Let's use the name cards to double-check.

When students are very familiar with this routine, with the total number in their class, and with making and comparing groups, you can pose a more difficult problem. For example:

If we know four students are NOT HERE in school today, how many students are HERE today? What are all the ways we can figure that out, without counting off?

Some students might suggest breaking four cubes off the attendance stick and counting the rest. Others might suggest counting back from the total number of students. Still others might suggest counting up from 4 to the total number of students.

In addition to being real data that students can see and relate to every day, attendance offers manageable numbers to work with. Repetition of this routine over the school year is important; only after students are familiar with the routine will they begin to focus on the numbers involved. Gradually, they will start to make some important connections between counting and comparing quantities.

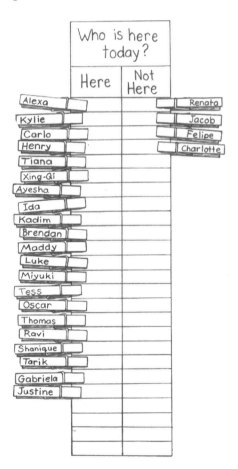

About Classroom Routines

Counting Jar

Counting is the foundation for much of the number work that students do in kindergarten and in the primary grades. Children learn to count by counting and hearing others count. Similarly, they learn about quantity through repeated experiences with organizing and counting sets of objects. The Counting Jar routine offers practice with all of these.

When students count sets of objects in the jar, they are practicing the counting sequence. As in the Attendance routine, they begin to see the need to develop strategies for counting, including ways to double-check and keep track of what they have counted. By recording the number of objects they have counted, students gain experience in representing quantity and conveying mathematical information to others. Creating a new equivalent set gives them not only another opportunity to count, but also a chance to compare the two amounts.

> Does my set have the same number as the set in the jar? How do I know?
>
> The jar has 8 and I have 7. I need 1 more because 8 is 1 more than 7.

As students work, they are developing a real sense of both numbers and quantities.

The Counting Jar routine is introduced in the first unit of the kindergarten curriculum, *Mathematical Thinking in Kindergarten*. The basic activity is described here, followed by suggested variations for use throughout the school year on a weekly basis.

Materials and Preparation

Obtain a clear plastic container, at least 6 inches tall and 4–5 inches in diameter. Fill it with a number of interesting objects that are uniform in size and not too small, such as golf or table tennis balls, small blocks or tiles, plastic animals, or walnuts in the shell. The total should be a number that is manageable for most students in your class; initially, 5 to 12 objects would be appropriate quantities.

Prepare a recording sheet on chart paper. At the top, write *The Counting Jar,* followed by the name of the material inside. Along the bottom, write a number line. Some students might use this number line to help them count objects or as a reference for writing numerals. Place each number in a box to clearly distinguish one from another.

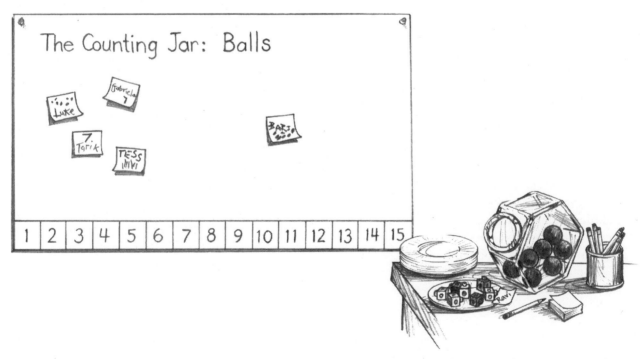

About Classroom Routines

Laminate this chart so that students can record their counts on the chart with stick-on notes, write-on/wipe-off markers, or small scraps of paper and tape; these can later be removed and the chart reused.

Also make available one paper plate for each student and sets of countable materials, such as cubes, buttons, keys, teddy bear counters, or color tiles, so that students can create a new set of materials that corresponds to the quantity in the Counting Jar.

Counting

How Many in the Jar? This routine has three basic steps:

- Working individually or in pairs, students count the objects in the Counting Jar.
- Students make a representation that shows how many objects are in the jar and place their response on the chart.
- Students count out another set of objects equivalent to the quantity in the Counting Jar. They place this new set on a paper plate, write their name on the plate, and display their equivalent collection near the Counting Jar.

As you use the Counting Jar throughout the school year, call attention to it in a whole-group meeting whenever you have changed the material or the amount inside the jar. Then leave it in a convenient location for two or three days so that everyone has a chance to count. After most students have counted individually, meet with the whole class and count the contents together.

Note: Some kindergarten teachers use a very similar activity for estimation practice. We exclude the task of estimation from the basic activity because until students have a sense of quantity, a sense of how much 6 is, a sense of what 10 balls look like compared to 10 cubes, it is difficult for them to estimate or predict how large a quantity is. When students are more familiar with the routine and have begun to develop a sense of quantity, you might include the variations suggested for estimation.

One More, One Less When students can count the materials in the jar with a certain amount of accuracy and understanding, try this variation for work with the ideas "one more than" and "one less than." As you offer the Counting Jar activity, ask students to create a set of objects with one more (or less) than the amount in the jar.

Filling the Jar Ourselves When the Counting Jar routine is firmly established, give individuals or pairs of students the responsibility for filling the jar. Discuss with them an appropriate quantity to put in the jar or suggest a target number, and let students decide on suitable objects to put in the jar.

At-Home Counting Jars Suggest to families that they set up a Counting Jar at home. Offer suggestions for different materials and appropriate quantities. Family members can take turns putting sets of objects in the jar for others to count.

Estimation

Is It More Than 5? To introduce the idea of estimation, show students a set of five objects identical to those in the Counting Jar. This gives students a concrete amount for reference to base their estimate on. As they look at the known quantity, ask them to think about whether there are *more than* five objects in the jar. The number in the reference group can grow as the number of objects in the jar changes, and you can begin to ask "Is the amount in the jar more than 8? more than 10?"

More or Less Than Yesterday? You can also encourage students to develop estimation skills when the material in the jar stays the same over several days but the quantity changes. In this situation, students can use reasoning like this:

Last time, when there were 8 blocks in the jar, it was filled up to *here*. Now it's a little higher, so I think there are 10 or 11 blocks.

Appendix: About Classroom Routines

About Classroom Routines

Calendar

"Calendar," with its many rituals and routines, is a familiar kindergarten activity. Perhaps the most important idea, particularly for young students, is viewing the calendar as a real-world tool that we use to keep track of time and events. As students work with the calendar, they become more familiar with the sequence of days, weeks, and months, and relationships among these periods of time. Time and the passage of time are challenging ideas for most 5- and 6-year-olds, and the ideas need to be linked to their own direct experiences. For example, explaining that an event will occur *after* a child's birthday or *before* a familiar holiday will help place that event in time for them.

The Calendar routine is introduced in the first unit of the kindergarten curriculum, *Mathematical Thinking in Kindergarten*. The basic activity is described here, followed by suggested variations for daily use throughout the school year.

Materials and Preparation

In most kindergarten classrooms, a monthly calendar is displayed where everyone can see it when the class gathers as a whole group. A calendar with date cards that can be removed or rearranged allows for greater flexibility than one without. Teachers make different choices about how to display numbers on this calendar. We recommend displaying all the days, from 1 to 30 or 31, all month long. This way the sequence of numbers and the total number of days are always visible, thus giving students a sense of the month as a whole.

You can use stick-on labels to highlight special days such as birthdays, class trips or events, non-school days, or holidays. Similarly, find some way to identify *today* on the calendar. Some teachers have a special star or symbol to clip on today's date card, or a special tag, much like a picture frame, that hangs over today's date.

A Sense of Time

The Monthly Calendar When first introducing the calendar, ask students what they notice. They are likely to mention a wide variety of things, including the colors they see on the calendar, pictures, numbers, words, how the calendar is arranged, and any special events they know are in that particular month. If no one brings it up, ask students what calendars are for and how we use them.

At the beginning of each month, involve students in organizing the dates and recording special events on the calendar. The following questions help them understand the calendar as a tool for keeping track of events in time:

If our trip to the zoo is on the 13th, on which day should we hang the picture of a lion?

Is our trip tomorrow? the next day? this week?

What day of the week will we go to the zoo?

How Much Longer? Many students eagerly anticipate upcoming events or special days. Ask students to figure out how much longer it is until something, or how many days have passed since something happened. For example:

How many more days is it until Alexa's birthday?

Today is November 4. How many more days is it until November 10?

How many days until the end of the month?

How many days have gone by since our parent breakfast?

Ask students to share their strategies for finding the number of days. Initially many students will

count each subsequent day. Later some students may begin to find answers by using their growing knowledge of calendar structure and number relationships:

> I knew there were three more days in this row, and I added them to the three in the next row. That's six more days.

Calculating "how many more days" on the calendar is not an easy task. Quite likely students will not agree on what days to count. Consider the following three good answers, all different, to this teacher's question:

Today is October 4. Ida's birthday is on October 8. How many more days until her birthday?

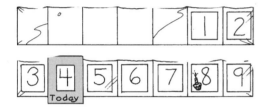

Tess: I think there are four more days because it's 4 . . . *[counting on her fingers]* 5, 6, 7, 8.

Ravi: There are three more days. See? *[He points to the three calendar dates between October 4 and October 8—5, 6, and 7—and counts three date cards.]*

Gabriela: It's five more days until her birthday. *[Using the calendar, she points to today and counts "1, 2, 3, 4 , 5," ending on October 8.]*

All of these students made sense of their answers and, considering their reasoning, all three were correct. That's why, when asking "how many more?" questions based on the calendar, it is important also to ask students to explain their thinking.

Numbers on the Calendar

Counting Days The calendar is a place where students can daily visit and become more familiar with the sequence of counting numbers up to 31. Because the numbers on the calendar represent the number of days in a month, the calendar is actually a way of *counting days*. You can help students with this idea:

Today is September 13. Thirteen days have already gone by in this month. If we start counting on 1, what number do you think we will end up on? Let's try it.

As you involve students in this way, they have another chance to see that numbers represent a quantity, in this case a number of days.

Missing Numbers or Mixed-Up Numbers Once students are familiar with the structure of the calendar and the sequence of numbers, you can play two games that involve removing and rearranging the dates. To play Missing Numbers, choose two or three dates on the monthly calendar and either remove or cover them. As students guess which numbers are missing, encourage them to explain their thinking and reasoning. Do they count from the number 1 or do they count on from another number? Do they know that 13 comes *after* 12 and *before* 14?

Mixed-Up Numbers is played by changing the position of numbers on the calendar so that some are out of order. Students then fix the calendar by pointing out which numbers are out of order.

Patterns on the Calendar

Looking for Patterns Some teachers like to point out patterns on the calendar. The repeating sequence of the days of the week and the months of the year are patterns that help students explore the cyclical nature of time. Many students quickly recognize the sequence of numbers 1 to 30 or 31, and some even recognize another important pattern on the calendar: that the columns increase by 7. However, in order to maintain the focus on the calendar as a tool for keeping track of time, we recommend using the Calendar routine only to note patterns that exist within the structure of the calendar and the sequence of days and numbers. The familiar activity of adding pictures or shapes to form repeating patterns can be better done in another routine, Patterns on the Pocket Chart.

About Classroom Routines

Today's Question

Collecting, representing, and interpreting information are ongoing activities in our daily lives. In today's world, organizing and interpreting data are vital to understanding events and making decisions based on this understanding. Because young students are natural collectors of materials and information, working with data builds on their natural curiosity about the world and people.

Today's Question offers students regular opportunities to collect information, record it on a class chart, and then discuss what it means. While engaged in this data collection and analysis, students are also counting real, meaningful quantities (How many of us have a pet?) and comparing quantities that are significant to them (Are there more girls in our class or more boys?). When working with questions that have only two responses, students explore part-whole relationships as they consider the total number of answers from the class and how that amount is broken into two parts.

Today's Question is introduced in the first unit of the kindergarten curriculum, *Mathematical Thinking in Kindergarten*. The basic routine is described here, followed by variations. Plan to use this routine throughout the school year on a weekly basis, or whenever a suitable and interesting question arises in your classroom.

Materials and Preparation

Prepare a chart for collecting students' responses to Today's Question. If you plan to use this routine frequently, either laminate a chart so that students can respond with wipe-off markers, or set up a blank chart on 11-by-17-inch paper and make multiple photocopies. The drawback of a laminated wipe-off chart is that you cannot save the information collected; with multiple charts, you can look back at data you have collected earlier or compare data from previous questions.

Make a section across the top of the chart, large enough to write the words *Today's Question* followed by the actual question being asked.

Mark the rest of the chart into two equal columns (later, you may want three columns). Leave enough space at the top of each column for the response choices, including words and possibly a sketch as a visual reminder.

Leave the bottom section (the largest part of the chart) blank for students to write their names to indicate their response. Your chart will look something like this:

Later in the year, you may want a chart with write-on lines in the bottom section to help students to compare numbers of responses in the two or three categories. Be sure to allow one line for each child in the class. Lines are also helpful guides if you collect data with "name pins," or clothespins marked on both sides with student names, as suggested for the Attendance routine (see illustration on p. 99).

Choosing Questions

Especially during the first half of the school year, try to choose questions with only two responses. With two categories of data, students are more likely to see the part-whole relationship between the number of responses in each category and the total number of students in the class.

About Classroom Routines

As your students become familiar with the routine and with analyzing the data they collect, you may decide to add a third response category. This is useful for questions that might not always elicit a clear yes-or-no response, such as these:

Do you think it will rain? *(yes, no, maybe)*

Do you want to play outside today? *(yes, no, I'm not sure)*

Do you eat lunch at school? *(yes, no, sometimes)*

As you choose questions and set up the charts for this routine, consider the full range of responses and modify or drop the question if there seem to be too many possible answers. Later in the year, as students become familiar with this routine, you may want to involve them in organizing and choosing Today's Question.

Questions About the Class With Today's Question, students can collect information about a group of people and learn more about their classmates. For example:

Are you a boy or a girl?

Are you 5 or 6 years old?

Do you have a younger brother?

Do you have a pet?

Did you bring your lunch to school today?

Do you go to an after-school program?

Do you like ice cream?

Did you walk or ride to school this morning?

Some teachers avoid questions about potentially sensitive issues (Have you lost a tooth? Can you tie your shoes?), while other use this routine to carefully raise some of these issues. Whichever you decide, it is best to avoid questions about material possessions (Does your family have a computer?).

Questions for Daily Decisions When you pose questions that involve students in making decisions about their classroom, they begin to see that they are collecting real data for a purpose. These data collection experiences underscore one of the main reasons for collecting data in the real world: to help people make decisions. For example:

Which book would you like me to read at story time? (Display two books.)

Would you prefer apples or grapes for snack?

Should we play on the playground or walk to the park today?

Questions for Curriculum Planning Some teachers use this routine to gather information that helps them plan the direction of a new curriculum topic or lesson. For example, you can learn about students' previous experiences and better prepare them before reading a particular story, meeting a special visitor, or going on a field trip, with questions like these:

Have you ever read or heard this story?

Have you ever been to the science museum?

Have you ever heard of George Washington?

For questions of this type, you might want to add a third possible response (*I'm not sure* or *I don't know*).

Discussing the Data

Data collection does not end with the creation of a representation or graph to show everyone's responses. In fact, much of the real work in data analysis begins after the data has been organized and represented. Each time students respond to Today's Question, it is important to discuss the results. Consider the following questions to promote data analysis in classroom discussions:

What do you think this graph is about?

What do you notice about this graph?

What can you tell about [the favorite part of our lunch] by looking at this graph?

If we went to another classroom, collected this same information, and made a graph, do you think that graph would look the same as or different from ours?

Graphs and other visual representations of the data are vehicles for communication. Thinking about what a graph represents or what it is communicating is a part of data analysis that even the youngest students can and should be doing.

About Classroom Routines

Patterns on the Pocket Chart

Mathematics is sometimes called "the science of patterns." We often use the language of mathematics to describe and predict numerical or geometrical regularities. When young students examine patterns, they look for relationships among the pattern elements and explore how that information can be used to predict what comes next. The classroom routine Patterns on the Pocket Chart offers students repeated opportunities to describe, copy, extend, create, and make predictions about repeating patterns. The use of a 10-by-10 pocket chart to investigate patterns of color and shape builds a foundation for the later grades, when this same pocket chart will display the numbers 1 to 100 and students will investigate patterns in the arrangement of numbers.

This routine is introduced in the second unit of the kindergarten curriculum, *Pattern Trains and Hopscotch Paths.* The basic routine is described here, followed by variations for use throughout the school year on a weekly basis.

Materials and Preparation

For this routine you will need a pocket chart, such as the vinyl Hundred Number Wall Chart (with transparent pockets and removable number cards). You will also need 2-inch squares of construction paper of different colors, a set of color tiles (ideally, the colors of the paper squares will match the tiles), a set of 20–30 What Comes Next? cards. These cards, with a large question mark in the center, are cut slightly larger than 2 inches so they will cover the colored squares. A blackline master for these cards is provided in the unit *Pattern Trains and Hopscotch Paths.* You can easily make your own cards with tagboard and a marking pen.

For the variation Shapes, Shells, and Such, you can use math manipulatives such as pattern blocks and interlocking cubes, picture or shape cards, or collections of small objects, such as buttons, keys, or shells. The only limitation is the size of the pockets on your chart.

What Comes Next?

Before introducing this activity, arrange an a-b repeating pattern in the first row of the pocket chart using ten paper squares in two colors of your choice. Beginning with the fifth position, cover each colored square with a What Comes Next? (question mark) card.

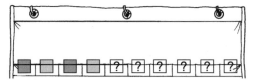

Gather students where the pocket chart is clearly visible and they have a place to work with color tiles, either on the floor or at tables.

Begin by asking students what they notice about the chart. Some may comment on the structure of the chart, some on the two-color pattern, and others may notice the question marks. Explain that each time they see one of these question marks, they should think "What comes next?" and decide what color might be under that card.

Provide each pair with a small cup of color tiles that match the paper squares. Ask students to build the first part of the pattern with color tiles and then predict what color comes next.

Who can predict, or guess, what color is hidden under each question mark on our chart? Use the tiles in your cup to show me what color would come next. How do you know?

Now, with your partner, see if you can make this pattern longer, using the tiles in your cup. Stop when your pattern has ten tiles.

When everyone has made a longer pattern, "read" the pattern together as a whole class. Verbalizing the pattern they are considering often helps students internalize it, recognize any errors in the pattern, and determine what comes next.

This basic activity can be done quickly, especially if students do not build the pattern with tiles. Many teachers integrate this routine into their

108 ■ Appendix: About Classroom Routines

About Classroom Routines

group meeting time on a regular basis, making one or two patterns on the pocket chart and asking students to predict what comes next.

Initially, use only two colors or two variables in the patterns. In addition to a-b (for example, blue-green) repeating patterns, build two-color patterns such as a-a-b (blue-blue-green), a-b-b (blue-green-green), or a-a-b-b (blue-blue-green-green).

Variations

Making Longer Patterns When students are familiar with the basic activity, they can investigate what happens to an a-b pattern when it "wraps around" and continues to the next line. If the pattern continues in a left-to-right progression, the pattern that emerges is the same one older students see when they investigate the patterns of odd and even numbers on the 100 chart.

Shapes, Shells, and Such Color is just one variable for patterns; others can be made using a wide variety of materials and pictures.

shell, shell, button, shell, shell, button

triangle, square, triangle, square

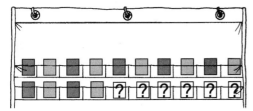

Picture cards, sometimes used by kindergarten teachers to make patterns on the calendar, make great patterns on the pocket chart without the distraction of the calendar elements.

What Comes *Here*? Predicting what comes next is an important idea in learning about patterns. Also important is being able to look ahead and predict what comes *here?* even further down the line. Instead of asking for the *next* color in a pattern sequence, point to a pocket three or four squares along and ask students to predict the color under that question mark. As you collect responses, ask students to explain how they predicted that color.

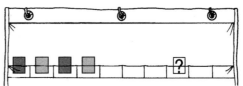

Border Patterns Explore a repeating pattern that extends around the entire outer edge of the pocket chart. Begin by filling the top row of the chart and asking what color would come next if this pattern turned the corner and went down the right side of the chart. Continue adding squares to finish the border. Every few days, begin a new pattern and ask students to help you complete the border. Start with a-b patterns. Gradually vary the pattern type, but continue to use only two colors, trying patterns such as a-a-b, a-b-b, a-a-b-b, or a-a-b-a. Ask students to notice which types make a continuous pattern all around the border and which do not.

With any border pattern, you can include a few What Comes Next? cards and ask students to predict the color of a particular pocket.

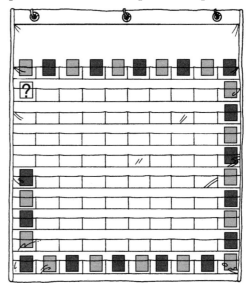

Patterns for Choice Time Hang the pocket chart where students can reach it. During free time or Choice Time, two or three students can work together to make their own pattern on the pocket chart, using colored paper squares or color tiles.

Appendix: About Classroom Routines ■ **109**

TIPS FOR THE LINGUISTICALLY DIVERSE CLASSROOM

It is likely that more students with limited English proficiency will be enrolled in kindergarten than any other grade. Moreover, many will be at the earliest stages of language acquisition. By correctly identifying a student's current level of English, you can create appropriate stimuli to ensure successful communication when presenting activities from *Investigations*.

The four stages of language acquisition are characterized as follows:

- **Preproduction** Students do not speak the language at this stage; they are dependent upon modeling, visual aids, and context clues to obtain meaning.
- **Early production** Students begin to produce isolated words in response to comprehensible questions. Responses are usually *yes*, *no*, or other single-word answers.
- **Speech emergence** Students now have a limited vocabulary and can respond in short phrases or sentences. Grammatical errors are common.
- **Intermediate fluency** Students can engage in conversation, producing full sentences.

You need to be aware of these four levels of proficiency while applying the following tips. The goal is always to ensure that students with limited English proficiency develop the same understandings as their English-speaking peers as they participate in this unit.

Tips for Small-Group Work Whenever possible, pair students with the same linguistic background and encourage them to complete the task in their native language. Students are more likely to have a successful exchange of ideas when they speak the same language. In other situations, teach all students how to make their communications comprehensible. For example, encourage students to point to objects they are discussing.

Tips for Whole-Class Activities To keep whole-group discussions comprehensible, draw simple sketches or diagrams on the board to illustrate key words; point to objects being discussed; use contrasting examples to help explain the attribute under discussion; model all directions; choose students to model activities or act out scenarios.

Tips for Observing the Students Assessment in the kindergarten units is based on your observations of students as they work, either independently or in groups. At times you will intervene by asking questions to help you evaluate a student's understanding. When questioning students, it is crucial not to misinterpret responses that are incomplete simply because of linguistic difficulties.

In many cases, students may understand the mathematical concept being asked about but not be able to articulate their thoughts in English. You need to formulate questions that allow students to respond at their stage of language acquisition in a way that indicates their mathematical understanding.

For example, this unit includes this suggestion for observing the class: "Are students able to use the vocabulary *longer than* and *shorter than* as they compare objects to their towers?" When you are observing students at the speech-emergence and intermediate-fluency stages of acquisition, you will likely hear such words spoken. However, students at earlier stages will probably not use those English terms. Therefore, you need to base your assessment on less-verbal indicators.

With students at the preproduction stage, for example, consider an alternative that calls for a nonverbal response: "Can the students show you something that is shorter than their tower?" With students at the early-production stage, you might look for single-word responses: "Can students tell you whether a particular object is *shorter* or *longer* than their tower?"

As you observe the students working, keep in mind which guidelines are appropriate for students at the different stages of language acquisition. Following is a categorization of typical questions from this unit.

Questions appropriate for students at the preproduction stage:

- How comfortable are students creating the groups of objects for each [counting book] page? Do they show more than one group for each page? If they are illustrating the "2" page, do all the groups have two things in them? Do students color in an accurate number of squares?
- How do students go about creating a group with a particular number of objects? Do they create sets of objects that have the same amount as the jar? Do they double-check their counts? Do they compare their new set to the set of objects in the jar?
- How do they position their tower while comparing it to an object? Do they stand items next to each other on the floor or table to compare them? If they lay the tower and object side by side, do they align them at one end?
- What dimensions of objects do students choose to measure? How do they determine which dimension is the longest? Do they ever measure two different dimensions of the same object?
- How do students arrange the tiles? Do the arrangements seem random?

Questions appropriate for students at the early-production and early-speech-emergence stages:

- How accurate are students as they count? Do they say one number for each object? Do they count each object once, or do they count some more than once or skip some?
- Can students count out loud to 6 successfully? Do they forget certain number names? Which ones? Do they mix the order of the number names?

Questions appropriate for students at the late-speech-emergence and intermediate-fluency stages:

- Are students able to use the vocabulary *longer than* and *shorter than* as they compare objects to their towers?
- Can they explain why or how they chose where to place a name or tower? What reasoning do they use?

- How do students describe and compare their towers? Do they use comparative terms such as *most* and *least*?
- How do students think about and describe their arrangements? What kinds of strategies do they have for remembering them? Are their strategies primarily visual? color-related? numerical? a combination of these?
- Can students explain how they arranged the tiles?

VOCABULARY SUPPORT FOR SECOND-LANGUAGE LEARNERS

The following activities will help ensure that this unit is comprehensible to students who are acquiring English as a second language. The suggested approach is based on *The Natural Approach: Language Acquisition in the Classroom* by Stephen D. Krashen and Tracy D. Terrell (Alemany Press, 1983). The intent is for second-language learners to acquire new vocabulary in an active, meaningful context.

Note that *acquiring* a word is different from *learning* a word. Depending on their level of proficiency, students may be able to comprehend a word upon hearing it during an investigation, without being able to say it. Other students may be able to use the word orally, but not read or write it. The goal is to help students naturally acquire targeted vocabulary at their present level of proficiency.

more, most, less, least, fewer, fewest

1. Ask four students to come forward. Hand out a collection of pencils, giving one to the first student, four to the next two students, and about eight to the last student. Make a point of physically comparing handfuls as you pass them out, and use comparative terms to describe them.

   Luis has *more* pencils than Tuan.
   Mali and Luis have the *same* number of pencils.
   Tatiana has *more* pencils than anyone.
   Mali, Tuan, and Luis all have *fewer* pencils than Tatiana.
   Tuan has the *fewest* pencils. He has the *least*.

2. The four students now trade places with someone else in the class, handing over their pencils as the new four come forward. As they hold up their handfuls, ask students to point out who has the most, who has the least (or fewest), and the two who have the same number.

3. Hold six loose cubes in your hands. Ask volunteers to distribute more cubes from the tub to the group. Indicate who should get more than you, who should get fewer, and who should get the most and the least. Ask the rest of the group to verify that the cubes were correctly distributed.

longer, longest, shorter, shortest

1. Prepare three strips of paper for each student, one long, one medium, and one short. Place a similar set in front of you.

2. Use the strips to model commands that make these words comprehensible. For example:

   Take the longest strip and put it on your head.
   Put the shortest strip on your shoulder.
   Sit on the longest strip.
   Step on the shortest strip.

3. Hold up your shortest strip in the air like a flag, and tell students to hold up a strip that is longer than yours. Then hold up a longer strip and challenge them to hold up one that's shorter.

4. Continue the game, holding up strips of different lengths and challenging students to recognize both shorter and longer strips.

higher, taller

1. Put two or three blocks in a stack. Explain that you are going to make the stack higher, and add another block on top. Do the same with a second stack.

2. Call on volunteers to come up to each stack, adding blocks one by one to make a higher tower. Lead the group in a chant of "higher, higher, taller, taller," as individuals keep adding one more block.

3. Quickly make several other towers, of varying heights but shorter than the originals. Point to different pairs of towers and in each case, ask students to indicate which one is higher or taller than the other.

Blackline Masters

Family Letter   114

Investigation 1
The Counting Book (eight sheets)   115
Student Sheet 1, Grab and Count   123

Investigation 2
Student Sheet 2, My Inventory Bag   124

Investigation 3
How to Play Compare   125
Number Cards   126

Investigation 4
Student Sheet 3, Our Names   130
Paper Towers for Collect 10   132
Investigating Names at Home   140

Investigation 5
Racing Bears Gameboard   133
How to Play Collect 10 Together   131

Investigation 6
One-Inch Grid Paper   134

General Resources for the Unit
Cube Strips   135
Choice Board Art   136

_____, 19____

Dear Family,

Our class is beginning a mathematics unit called *Collecting, Counting, and Measuring*. We will be exploring numbers with a variety of activities: making counting books, taking inventory of classroom materials, counting and comparing handfuls of cubes, counting the number of letters in our names, and playing counting games. With many experiences like these over the year, the children build their knowledge of the counting sequence and come to recognize the amount each number represents.

Children often enter kindergarten thinking of counting as a string of words, like a song. Throughout this school year, they will gradually begin to use counting as a tool. By counting many things and listening to others counting, your child will develop a meaningful sense of numbers.

While our class is working on this unit, you might try some of the following:

- We play several versions of a game called Grab and Count. The children take a handful of some material, then count to find out how many items they were able to grab. Generally, they are counting amounts up to 10 or 12. You might do this with things you have at home, such as checkers, small blocks, or spools of thread. Avoid buttons, coins, or other things that are too small, or your child may grab more than he or she is able to count. If you get handfuls that seem too large, you can count them out loud together so your child can hear and practice the sequence. Sometimes, you might both grab a handful, then count them. Who grabbed more? Who grabbed less?

- When we "take inventory" of materials in our classroom, you and your child might take a similar inventory at home. For example, count the number of cereal bowls, beds, chairs, people, or pillows. Encourage your child to find ways to keep track of which ones have been counted and which still need to be counted.

- One of our class activities will be counting letters in our names. You can do the same at home with the names of family members or a group of friends. Which name has the most letters? Which has the least? Can your child put the names in order by length?

Thank you for sharing in our kindergarten mathematics activities.

Sincerely,

_____'s
Counting Book

0

116

1

2

3

4

5

6

Name _____ Date _____

Student Sheet 1

Grab and Count

Name _____ Date _____

Student Sheet 2

My Inventory Bag

Investigation 2
Collecting, Counting, and Measuring

HOW TO PLAY COMPARE

Note to Families: This is a variation on a number game we play in class. Use this sheet to review the directions with your child. When you play the game together at home, give your child lots of time to think about the numbers on the cards. Please keep the game directions and the Number Cards in a safe place at home for continued use.

Materials: Deck of Number Cards 0–6 (remove the 7–10 cards and the wild cards from the complete set)

Counters (about 15, optional)

Players: 2

Object: Decide which of two cards shows a larger number.

How to Play

1. Mix the Number Cards and deal them evenly to each player. Both players place their stack of cards facedown in front of them.

2. At the same time, both players turn over the top card in their stack. Look at the numbers. (You can use counters to show the numbers if you like.) If your number is larger, you say "Me!" If the two cards are the same, turn over the next card.

3. Keep turning over cards. Each time, say "Me!" if your number is larger.

4. The game is over when you have both turned over all the cards in your stack.

Variations

Try some of these different ways to play the game.

- If you have the **smaller** number, you say "Me."

- Play with three people. Look at all three numbers. If you have the largest number, say "Me."

- Add four wild cards to the deck. If you turn over a wild card, you can decide what number it is.

NUMBER CARDS (page 1 of 4)

0	0	0	0
1	1	1	1
2	2	2	2

Investigation 3
Collecting, Counting, and Measuring

NUMBER CARDS (page 2 of 4)

3	3	3	3
4	4	4	4
5	5	5	5

Investigation 3
Collecting, Counting, and Measuring

NUMBER CARDS (page 3 of 4)

NUMBER CARDS (page 4 of 4)

9	9	9	9
10	10	10	10
Wild Card	Wild Card	Wild Card	Wild Card

Name _____ Date _____

Student Sheet 3

Our Names

My name is _____

There are _____ letters in my name.

Shorter	Same	Longer

Investigation 4
Collecting, Counting, and Measuring

HOW TO PLAY COLLECT 10 TOGETHER

Note to Families: This is a game we have been playing in class. Use this sheet to review the directions with your child. When you play it together at home, encourage your child to do the counting for both of you. Please keep both the game directions and the Number Cards in a safe place at home for continued use.

Materials: Number Cards 1–3 (from the full set of Number Cards, use only the twelve cards for the numbers 1, 2, and 3)

15–20 counters, such as buttons or beans

Players: 2

Object: With a partner, collect 10 counters.

How to Play

1. Mix the cards and stack them facedown. To start, one player turns over a card. What number is it? Take that many counters.

2. Now the other player turns over a new card. Take that many counters and add them to the collection.

3. Take turns. After each turn, check the total number of counters in your collection. The game ends when you have 10 counters.

Variations

Try some of these different ways to play the game.

- Change the number of counters you must collect. For example, play Collect 8 Together or Collect 12 Together.

- Play with three people.

You can use this Tower of 10 to help you keep track of how many counters you have collected so far.

PAPER TOWERS FOR COLLECT 10

Investigation 4
Collecting, Counting, and Measuring

RACING BEARS GAMEBOARD

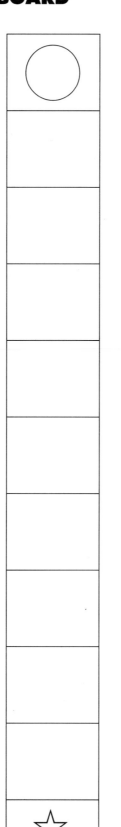
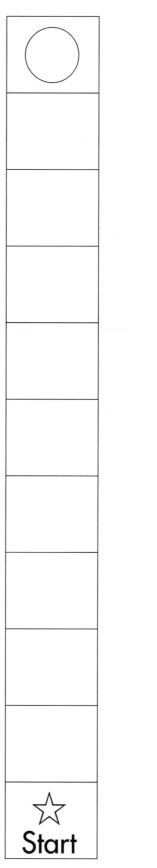

133

Investigation 5
Collecting, Counting, and Measuring

ONE-INCH GRID PAPER

CUBE STRIPS

CHOICE BOARD ART (page 1 of 4)

Choice Board art for
My Counting Book

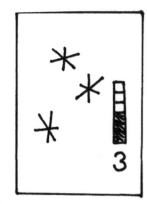

Choice Board art for
Grab and Count

Choice Board art for
Counting Jar

Choice Board art for
Inventory Bags

General Resource
Collecting, Counting, and Measuring

CHOICE BOARD ART (page 2 of 4)

Choice Board art for
Measuring Table

Choice Board art for
Grab and Count: Which Has More?

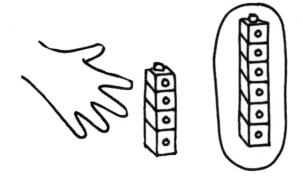

Choice Board art for
Compare

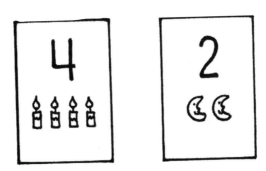

CHOICE BOARD ART (page 3 of 4)

Choice Board art for
Comparing Names

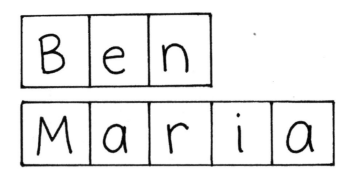

Choice Board art for
Grab and Count: Compare

Choice Board art for
Collect 10 Together

CHOICE BOARD ART (page 4 of 4)

Choice Board art for
Grab and Count: Least to Most

Choice Board art for
Racing Bears

Choice Board art for
Books of Six

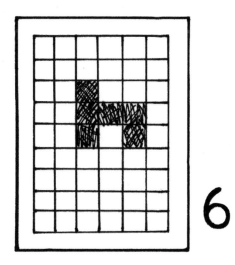

INVESTIGATING NAMES AT HOME

Dear Family,

In class, we have been counting the number of letters in our names. We are also comparing all our names to find which ones are longest and which ones are shortest.

Help your child do the same thing with the first names of the people in your family. You can help your child print the names below. Or, you might print the names on another sheet of paper so your child can copy them below.

Then watch your child count the letters in each name. Which is the longest name in your family? Which is the shortest?

Your child might be interested in putting all the family names in order, from longest to shortest, or shortest to longest.

It will be fun to compare all the names the children bring to class!

_____

Names in my family